M000289983

*The Elements of Integration
and Lebesgue Measure*

The Elements of Integration
and Lebesgue Measure

ROBERT G. BARTLE

Eastern Michigan University
and
University of Illinois

Wiley Classics Library Edition Published 1995

A Wiley-Interscience Publication
JOHN WILEY & SONS, INC.
New York • Chichester • Brisbane • Toronto • Singapore

This text is printed on acid-free paper.

Copyright © 1966 by John Wiley & Sons, Inc.

Wiley Classics Library Edition Published 1995.

All rights reserved. Published simultaneously in Canada.

Reproduction or translation of any part of this work beyond
that permitted by Section 107 or 108 of the 1976 United
States Copyright Act without the permission of the copyright
owner is unlawful. Requests for permission or further
information should be addressed to the Permissions Department,
John Wiley & Sons, Inc., 605 Third Avenue, New York, NY
10158-0012.

Library of Congress Cataloging in Publication Data:

Library of Congress Catalog Card Number 66-21048

ISBN 0-471-04222-6

Printed in the United States of America

20 19 18 17 16 15 14 13 12 11

Preface

This book consists of two separate, but closely related, parts. The first part (Chapters 1-10) is subtitled *The Elements of Integration;* the second part (Chapters 11-17) is subtitled *The Elements of Lebesgue measure.* It is possible to read these two parts in either order, with only a bit of repetition.

The Elements of Integration is essentially a corrected reprint of a book with that title, originally published in 1966, designed to present the chief results of the Lebesgue theory of integration to a reader having only a modest mathematical background. This book developed from my lectures at the University of Illinois, Urbana-Champaign, and it was subsequently used there and elsewhere with considerable success. Its only prerequisites are a understanding of elementary real analysis and the ability to comprehend "ε-δ arguments". We suppose that the reader has some familarity with the Riemann integral so that it is not necessary to provide motivation and detailed discussion, but we do not assume that the reader has a mastery of the subtleties of that theory. A solid course in "advanced calculus", an understanding of the first third of my book *The Elements of Real Analysis*, or of most of my book *Introduction to Real Analysis* with D. R. Sherbert provides an adequate background. In preparing this new edition, I have seized the opportunity to correct certain errors, but I have resisted the temptation to insert additional material, since I believe that one of the features of this book that is most appreciated is its brevity.

The Elements of Lebesgue Measure is descended from class notes written to acquaint the reader with the theory of Lebesgue measure in the space R^p. While it is easy to find good treatments of the case $p = 1$, the case $p > 1$ is not quite as simple and is much less frequently discussed. The main ideas of Lebesgue measure are presented in detail in Chapters 10-15, although some relatively easy remarks are left to the reader as exercises. The final two chapters venture into the topic of nonmeasurable sets and round out the subject.

There are many expositions of the Lebesgue integral from various points of view, but I believe that the abstract measure space approach used here strikes directly towards the most important results: the convergence theorems. Further, this approach is particularly well-suited for students of probability and statistics, as well as students of analysis. Since the book is intended as an introduction, I do not follow all of the avenues that are encountered. However, I take pains not to attain brevity by leaving out important details, or assigning them to the reader.

Readers who complete this book are certainly not through, but if this book helps to speed them on their way, it has accomplished its purpose. In the References, I give some books that I believe readers can profitably explore, as well as works cited in the body of the text.

I am indebted to a number of colleagues, past and present, for their comments and suggestions; I particularly wish to mention N. T. Hamilton, G. H. Orland, C. W. Mullins, A. L. Peressini, and J. J. Uhl, Jr. I also wish to thank Professor Roy O. Davies of Leicester University for pointing out a number of errors and possible improvements.

ROBERT G. BARTLE

Ypsilanti and Urbana
November 20, 1994

Contents

The Elements of Integration

1. *Introduction*   1

Reasons for the development of the Lebesgue integral,
comparison with the Riemann integral, the extended real
number system

2. *Measurable Functions*   6

Measurable sets and functions, combinations, complex-
valued functions, functions between measurable spaces

3. *Measures*   19

Measures, measure spaces, almost everywhere, charges

4. *The Integral*   27

Simple functions and their integrals, the integral of a nonnegative extended real-valued measurable function, the
Monotone Convergence Theorem, Fatou's Lemma, properties of the integral

5. *Integrable Functions*   **41**

Integrable real-valued functions, positivity and linearity
of the integral, the Lebesgue Dominated Convergence
Theorem, integrands that depend on a parameter

6. *The Lebesgue Spaces L_p*   **52**

Normed linear spaces, the L_p spaces, Hölder's Inequality,
Minkowski's Inequality, the Completeness Theorem, the
space L_∞

7. *Modes of Convergence*   **65**

Relations between convergence in mean, uniform convergence, almost everywhere convergence, convergence in
measure, and almost uniform convergence, Egoroff's Theorem, the Vitali Convergence Theorem

8. *Decomposition of Measures*   **80**

Hahn and Jordan Decomposition Theorems, the Radon-
Nikodým Theorem, Lebesgue Decomposition Theorem,
the Riesz Representation Theorem for L_p

9. *Generation of Measures*   **96**

Measures on algebras of sets, the extension of measures,
Carathéodory and Hahn Extension Theorems, Lebesgue
and Lebesgue-Stieltjes measure, the Riesz Representation
Theorem for C

10. *Product Measures*   **113**

Rectangles, the Product Measure Theorem, sections, the
Monotone Class Lemma, Tonelli and Fubini Theorems

The Elements of Lebesgue Measure

11. *Volumes of Cells and Intervals*   **127**

Cells, intervals, length, cells in R^p, p-dimensional volume, translation invariance

12. *The Outer Measure*   **131**

The outer measure in R^p, properties of m^*, translation invariance

13. *Measurable Sets*   **139**

σ-Algebras, measure on a σ-algebra, the Carathéodory condition, Carathéodory's theorem, Lebesgue sets and Lebesgue measure, uniqueness of Lebesgue measure, some useful properties

14. *Examples of Measurable Sets*   **149**

Borel sets, null sets, translation invariance, existence of non-Borel sets

15. *Approximation of Measurable Sets*   **155**

Approximation by open sets, approximation by closed sets, approximation by compact sets, approximation by cells

16. *Additivity and Nonadditivity*   **161**

Additivity, Carathéodory revisited, inner measure

17. *Nonmeasurable and Non-Borel Sets*   165

The difference set, rational equivalence, Vitali's set, nonadditive decompositions, an ubiquitous set, a singular
function, existence of non-Borel sets

References   175

Index   177

The Elements of Integration

CHAPTER 1

Introduction

The theory of integration has its ancient and honorable roots in the "method of exhaustion" that was invented by Eudoxos and greatly developed by Archimedes for the purpose of calculating the areas and volumes of geometric figures. The later work of Newton and Leibniz enabled this method to grow into a systematic tool for such calculations.

As this theory developed, it has become less concerned with applications to geometry and elementary mechanics, for which it is entirely adequate, and more concerned with purely analytic questions, for which the classical theory of integration is not always sufficient. Thus a present-day mathematician is apt to be interested in the convergence of orthogonal expansions, or in applications to differential equations or probability. For him the classical theory of integration which culminated in the Riemann integral has been largely replaced by the theory which has grown from the pioneering work of Henri Lebesgue at the beginning of this century. The reason for this is very simple: the powerful convergence theorems associated with the Lebesgue theory of integration lead to more general, more complete, and more elegant results than the Riemann integral admits.

Lebesgue's definition of the integral enlarges the collection of functions for which the integral is defined. Although this enlargement is useful in itself, its main virtue is that the theorems relating to the interchange of the limit and the integral are valid under less stringent assumptions than are required for the Riemann integral. Since one frequently needs to make such interchanges, the Lebesgue integral is more convenient to deal with than the Riemann integral. To exemplify these remarks, let the sequence (f_n) of functions be defined for $x > 0$ by $f_n(x) = e^{-nx}/\sqrt{x}$. It is readily seen that the (improper) Riemann integrals

$$I_n = \int_0^{+\infty} \frac{e^{-nx}}{\sqrt{x}}\, dx$$

exist and that $\lim_{n \to \infty} f_n(x) = 0$ for all $x > 0$. However, since $\lim_{x \to 0} f_n(x) = +\infty$ for each n, the convergence of the sequence is certainly not uniform for $x > 0$. Although it is hoped that the reader can supply the estimates required to show that $\lim I_n = 0$, we prefer to obtain this conclusion as an immediate consequence of the Lebesgue Dominated Convergence Theorem which will be proved later. As another example, consider the function F defined for $t > 0$ by the (improper) Riemann integral

$$F(t) = \int_0^{+\infty} x^2\, e^{-tx}\, dx.$$

With a little effort one can show that F is continuous and that its derivative exists and is given by

$$F'(t) = -\int_0^{+\infty} x^3\, e^{-tx}\, dx,$$

which is obtained by differentiating under the integral sign. Once again, this inference follows easily from the Lebesgue Dominated Convergence Theorem.

At the risk of oversimplification, we shall try to indicate the crucial difference between the Riemann and the Lebesgue definitions of the integral. Recall that an **interval** in the set R of real numbers is a set which has one of the following four forms:

$$[a, b] = \{x \in R : a \leqslant x \leqslant b\}, \qquad (a, b) = \{x \in R : a < x < b\},$$
$$[a, b) = \{x \in R : a \leqslant x < b\}, \qquad (a, b] = \{x \in R : a < x \leqslant b\}.$$

In each of these cases we refer to a and b as the **endpoints** and prescribe

$b - a$ as the **length** of the interval. Recall further that if E is a set, then the **characteristic function** of E is the function χ_E defined by

$$\begin{aligned} \chi_E(x) &= 1, &&\text{if } x \in E, \\ &= 0, &&\text{if } x \notin E. \end{aligned}$$

A **step function** is a function φ which is a finite linear combination of characteristic functions of intervals; thus

$$\varphi = \sum_{j=1}^{n} c_j \chi_{E_j}.$$

If the endpoints of the interval E_j are a_j, b_j, we define the **integral** of φ to be

$$\int \varphi = \sum_{j=1}^{n} c_j(b_j - a_j).$$

If f is a bounded function defined on an interval $[a, b]$ and if f is not too discontinuous, then the **Riemann integral** of f is defined to be the limit (in an appropriate sense) of the integrals of step functions which approximate f. In particular, the **lower Riemann integral** of f may be defined to be the supremum of the integrals of all step functions φ such that $\varphi(x) \leqslant f(x)$ for all x in $[a, b]$, and $\varphi(x) = 0$ for x not in $[a, b]$.

The Lebesgue integral can be obtained by a similar process, except that the collection of step functions is replaced by a larger class of functions. In somewhat more detail, the notion of length is generalized to a suitable collection X of subsets of $\boldsymbol{R}$. Once this is done, the step functions are replaced by **simple functions**, which are finite linear combinations of characteristic functions of sets belonging to X. If

$$\varphi = \sum_{j=1}^{n} c_j \chi_{E_j}$$

is such a simple function and if $\mu(E)$ denotes the "measure" or "generalized length" of the set E in X, we define the integral of φ to be

$$\int \varphi = \sum_{j=1}^{n} c_j \mu(E_j).$$

If f is a nonnegative function defined on $\boldsymbol{R}$ which is suitably restricted, we shall define the **(Lebesgue) integral** of f to be the supremum of the integrals of all simple functions φ such that $\varphi(x) \leqslant f(x)$ for all x in R. The integral can then be extended to certain functions that take both signs.

Although the generalization of the notion of length to certain sets in R which are not necessarily intervals has great interest, it was observed in 1915 by Maurice Fréchet that the convergence properties of the Lebesgue integral are valid in considerable generality. Indeed, let X be any set in which there is a collection X of subsets containing the empty set $\emptyset$ and X and closed under complementation and countable unions. Suppose that there is a nonnegative measure function μ defined on X such that $\mu(\emptyset) = 0$ and which is **countably additive** in the sense that

$$\mu\left(\bigcup_{j=1}^{\infty} E_j\right) = \sum_{j=1}^{\infty} \mu(E_j)$$

for each sequence (E_j) of sets in X which are mutually disjoint. In this case an integral can be defined for a suitable class of real-valued functions on X, and this integral possesses strong convergence properties.

As we have stressed, we are particularly interested in these convergence theorems. Therefore we wish to advance directly toward them in this abstract setting, since it is more general and, we believe, conceptually simpler than the special cases of integration on the line or in R^n. However, it does require that the reader temporarily accept the fact that interesting special cases are subsumed by the general theory. Specifically, it requires that he accept the assertion that there exists a countably additive measure function that extends the notion of the length of an interval. The proof of this assertion is in Chapter 9 and can be read after completing Chapter 3 by those for whom the suspense is too great.

In this introductory chapter we have attempted to provide motivation and to set the stage for the detailed discussion which follows. Some of our remarks here have been a bit vague and none of them has been proved. These defects will be remedied. However, since we shall have occasion to refer to the system of extended real numbers, we now append a brief description of this system.

In integration theory it is frequently convenient to adjoin the two symbols $-\infty$, $+\infty$ to the real number system $\boldsymbol{R}$. (It is stressed that these symbols are not real numbers.) We also introduce the convention that $-\infty < x < +\infty$ for any $x \in \boldsymbol{R}$. The collection $\bar{\boldsymbol{R}}$ consisting of the set $\boldsymbol{R} \cup \{-\infty, +\infty\}$ is called the **extended real number system**.

One reason we wish to consider $\bar{\boldsymbol{R}}$ is that it is convenient to say that the length of the real line is equal to $+\infty$. Another reason is that we will frequently be taking the supremum ($=$ least upper bound) of a set of real numbers. We know that a nonempty set A of real numbers which has an upper bound also has a supremum (in $\boldsymbol{R}$). If we define the supremum of a nonempty set which does not have an upper bound to be $+\infty$, then every nonempty subset of $\boldsymbol{R}$ (or $\bar{\boldsymbol{R}}$) has a unique supremum in $\bar{\boldsymbol{R}}$. Similarly, every nonempty subset of $\boldsymbol{R}$ (or $\bar{\boldsymbol{R}}$) has a unique infimum ($=$ greatest lower bound) in $\bar{\boldsymbol{R}}$. (Some authors introduce the conventions that $\inf \emptyset = +\infty$, $\sup \emptyset = -\infty$, but we shall not employ them.)

If (x_n) is a sequence of extended real numbers, we define the **limit superior** and the **limit inferior** of this sequence by

$$\limsup x_n = \inf_m \left(\sup_{n \geqslant m} x_n \right),$$
$$\liminf x_n = \sup_m \left(\inf_{n \geqslant m} x_n \right).$$

If the limit inferior and the limit superior are equal, then their value is called the **limit** of the sequence. It is clear that this agrees with the conventional definition when the sequence and the limit belong to $\boldsymbol{R}$.

Finally, we introduce the following algebraic operations between the symbols $\pm\infty$ and elements $x \in \boldsymbol{R}$:

$$
\begin{aligned}
(\pm\infty) + (\pm\infty) &= x + (\pm\infty) = (\pm\infty) + x = \pm\infty, \\
(\pm\infty)(\pm\infty) &= +\infty, \ (\pm\infty)(\mp\infty) = -\infty, \\
x(\pm\infty) = (\pm\infty)x &= \pm\infty \qquad \text{if } x > 0, \\
&= 0 \qquad \ \ \text{if } x = 0, \\
&= \mp\infty \qquad \text{if } x < 0.
\end{aligned}
$$

It should be noticed that we do not define $(+\infty) + (-\infty)$ or $(-\infty) + (+\infty)$, nor do we define quotients when the denominator is $\pm\infty$.

CHAPTER 2

Measurable Functions

In developing the Lebesgue integral we shall be concerned with classes of real-valued functions defined on a set X. In various applications the set X may be the unit interval $I = [0, 1]$ consisting of all real numbers x satisfying $0 \leqslant x \leqslant 1$; it may be the set $N = \{1, 2, 3, \ldots\}$ of natural numbers; it may be the entire real line R; it may be all of the plane; or it may be some other set. Since the development of the integral does not depend on the character of the underlying space X, we shall make no assumptions about its specific nature.

Given the set X, we single out a family X of subsets of X which are "well-behaved" in a certain technical sense. To be precise, we shall assume that this family contains the empty set $\emptyset$ and the entire set X, and that X is closed under complementation and countable unions.

2.1 DEFINITION. A family X of subsets of a set X is said to be a σ-**algebra** (or a σ-**field**) in case:

   (i) $\emptyset$, X belong to X.
   (ii) If A belongs to X, then the complement $\mathscr{C}(A) = X \setminus A$ belongs to X.
   (iii) If (A_n) is a sequence of sets in X, then the union $\bigcup_{n=1}^{\infty} A_n$ belongs to X.

An ordered pair (X, X) consisting of a set X and a σ-algebra X of subsets of X is called a **measurable space**. Any set in X is called an

6

X-measurable set, but when the σ-algebra X is fixed (as is generally the case), the set will usually be said to be **measurable**.

The reader will recall the rules of De Morgan:

(2.1) $\qquad \mathscr{C}\left(\bigcup_\alpha A_\alpha\right) = \bigcap_\alpha \mathscr{C}(A_\alpha), \qquad \mathscr{C}\left(\bigcap_\alpha A_\alpha\right) = \bigcup_\alpha \mathscr{C}(A_\alpha).$

It follows from these that the intersection of a sequence of sets in X also belongs to X.

We shall now give some examples of σ-algebras of subsets.

2.2 EXAMPLES. (a) Let X be any set and let X be the family of all subsets of X.

(b) Let X be the family consisting of precisely two subsets of X, namely $\emptyset$ and X.

(c) Let $X = \{1, 2, 3, \ldots\}$ be the set N of natural numbers and let X consist of the subsets

$$\emptyset, \quad \{1, 3, 5, \ldots\}, \quad \{2, 4, 6, \ldots\}, \quad X.$$

(d) Let X be an uncountable set and X be the collection of subsets which are either countable or have countable complements.

(e) If X_1 and X_2 are σ-algebras of subsets of X, let X_3 be the intersection of X_1 and X_2; that is, X_3 consists of all subsets of X which belong to both X_1 and X_2. It is readily checked that X_3 is a σ-algebra.

(f) Let A be a nonempty collection of subsets of X. We observe that there is a smallest σ-algebra of subsets of X containing A. To see this, observe that the family of all subsets of X is a σ-algebra containing A and the intersection of all the σ-algebras containing A is also a σ-algebra containing A. This smallest σ-algebra is sometimes called the σ-**algebra generated by** A.

(g) Let X be the set R of real numbers. The **Borel algebra** is the σ-algebra B generated by all open intervals (a, b) in R. Observe that the Borel algebra B is also the σ-algebra generated by all closed intervals $[a, b]$ in R. Any set in B is called a **Borel set**.

(h) Let X be the set $\bar{R}$ of extended real numbers. If E is a Borel subset of R, let

(2.2) $E_1 = E \cup \{-\infty\}, \qquad E_2 = E \cup \{+\infty\}, \qquad E_3 = E \cup \{-\infty, +\infty\},$

and let $\bar{B}$ be the collection of all sets E, E_1, E_2, E_3 as E varies over B. It is readily seen that $\bar{B}$ is a σ-algebra and it will be called the **extended Borel algebra**.

In the following, we shall consider a fixed measurable space (X, X).

2.3 DEFINITION. A function f on X to R is said to be **X-measurable** (or simply **measurable**) if for every real number α the set

(2.3) $$\{x \in X : f(x) > \alpha\}$$

belongs to X.

The next lemma shows that we could have modified the form of the sets in defining measurability.

2.4 LEMMA. *The following statements are equivalent for a function f on X to R:*

(a) *For every $\alpha \in R$, the set $A_\alpha = \{x \in X : f(x) > \alpha\}$ belongs to X.*
(b) *For every $\alpha \in R$, the set $B_\alpha = \{x \in X : f(x) \leqslant \alpha\}$ belongs to X.*
(c) *For every $\alpha \in R$, the set $C_\alpha = \{x \in X : f(x) \geqslant \alpha\}$ belongs to X.*
(d) *For every $\alpha \in R$, the set $D_\alpha = \{x \in X : f(x) < \alpha\}$ belongs to X.*

PROOF. Since B_α and A_α are complements of each other, statement (a) is equivalent to statement (b). Similarly, statements (c) and (d) are equivalent. If (a) holds, then $A_{\alpha - 1/n}$ belongs to X for each n and since

$$C_\alpha = \bigcap_{n=1}^{\infty} A_{\alpha - 1/n},$$

it follows that $C_\alpha \in X$. Hence (a) implies (c). Since

$$A_\alpha = \bigcup_{n=1}^{\infty} C_{\alpha + 1/n},$$

it follows that (c) implies (a).   Q.E.D.

2.5 EXAMPLES. (a) Any constant function is measurable. For, if $f(x) = c$ for all $x \in X$ and if $\alpha \geqslant c$, then

$$\{x \in X : f(x) > \alpha\} = \emptyset,$$

whereas if $\alpha < c$, then

$$\{x \in X : f(x) > \alpha\} = X.$$

(b) If $E \in X$, then the **characteristic function** χ_E, defined by

$$\chi_E(x) = 1, \quad x \in E,$$
$$= 0, \quad x \notin E,$$

is measurable. In fact, $\{x \in X : \chi_E(x) > \alpha\}$ is either X, E, or $\emptyset$.

(c) If X is the set R of real numbers, and X is the Borel algebra B, then any continuous function f on R to R is Borel measurable (that is, B-measurable). In fact, if f is continuous, then $\{x \in R : f(x) > \alpha\}$ is an open set in R and hence is the union of a sequence of open intervals. Therefore, it belongs to B.

(d) If $X = R$ and $X = B$, then any monotone function is Borel measurable. For, suppose that f is monotone increasing in the sense that $x \leqslant x'$ implies $f(x) \leqslant f(x')$. Then $\{x \in R : f(x) > \alpha\}$ consists of a half-line which is either of the form $\{x \in R : x > a\}$ or the form $\{x \in R : x \geqslant a\}$, or is R or $\emptyset$.

Certain simple algebraic combinations of measurable functions are measurable, as we shall now show.

2.6 LEMMA. *Let f and g be measurable real-valued functions and let c be a real number. Then the functions*

$$cf, \quad f^2, \quad f + g, \quad fg, \quad |f|,$$

are also measurable.

PROOF. (a) If $c = 0$, the statement is trivial. If $c > 0$, then

$$\{x \in X : cf(x) > \alpha\} = \{x \in X : f(x) > \alpha/c\} \in X.$$

The case $c < 0$ is handled similarly.

(b) If $\alpha < 0$, then $\{x \in X : (f(x))^2 > \alpha\} = X$; if $\alpha \geqslant 0$, then

$$\{x \in X : (f(x))^2 > \alpha\}$$
$$= \{x \in X : f(x) > \sqrt{\alpha}\} \cup \{x \in X : f(x) < -\sqrt{\alpha}\}.$$

(c) By hypothesis, if r is a rational number, then

$$S_r = \{x \in X : f(x) > r\} \cap \{x \in X : g(x) > \alpha - r\}$$

belongs to X. Since it is readily seen that

$$\{x \in X : (f + g)(x) > \alpha\} = \bigcup \{S_r : r \text{ rational}\},$$

it follows that $f + g$ is measurable.

(d) Since $fg = \frac{1}{4}[(f + g)^2 - (f - g)^2]$, it follows from parts (a), (b), and (c) that fg is measurable.

(e) If $\alpha < 0$, then $\{x \in X : |f(x)| > \alpha\} = X$, whereas if $\alpha \geq 0$, then

$$\{x \in X : |f(x)| > \alpha\} = \{x \in X : f(x) > \alpha\} \cup \{x \in X : f(x) < -\alpha\}.$$

Thus the function $|f|$ is measurable.   Q.E.D.

If f is any function on X to R, let f^+ and f^- be the nonnegative functions defined on X by

(2.4) $f^+(x) = \sup\{f(x), 0\}, \quad f^-(x) = \sup\{-f(x), 0\}.$

The function f^+ is called the **positive part** of f and f^- is called the **negative part** of f. It is clear that

(2.5) $f = f^+ - f^- \quad$ and $\quad |f| = f^+ + f^-$

and it follows from these identities that

(2.6) $f^+ = \frac{1}{2}(|f| + f), \quad f^- = \frac{1}{2}(|f| - f).$

In view of the preceding lemma we infer that f is measurable if and only if f^+ and f^- are measurable.

The preceding discussion pertained to real-valued functions defined on a measurable space. However, in dealing with sequences of measurable functions we often wish to form suprema, limits, etc., and it is technically convenient to allow the extended real numbers $-\infty$, $+\infty$ to be taken as values. Hence we wish to define measurability for extended real-valued functions and we do this exactly as in Definition 2.3.

2.7 DEFINITION. An extended real-valued function on X is X-**measurable** in case the set $\{x \in X : f(x) > \alpha\}$ belongs to X for each real number α. The collection of all extended real-valued X-measurable functions on X is denoted by $M(X, X)$.

Observe that if $f \in M(X, X)$, then

$$\{x \in X : f(x) = +\infty\} = \bigcap_{n=1}^{\infty} \{x \in X : f(x) > n\},$$

$$\{x \in X : f(x) = -\infty\} = \mathscr{C}\left[\bigcup_{n=1}^{\infty} \{x \in X : f(x) > -n\}\right],$$

so that both of these sets belong to X.

The following lemma is often useful in treating extended real-valued functions.

2.8 LEMMA.   *An extended real-valued function f is measurable if and only if the sets*

$$A = \{x \in X : f(x) = +\infty\}, \quad B = \{x \in X : f(x) = -\infty\}$$

belong to X and the real-valued function f_1 defined by

$$\begin{aligned} f_1(x) &= f(x), && \text{if } x \notin A \cup B, \\ &= 0, && \text{if } x \in A \cup B, \end{aligned}$$

is measurable.

PROOF.   If f is in $M(X, X)$, it has already been noted that A and B belong to X.   Let $\alpha \in R$ and $\alpha \geqslant 0$, then

$$\{x \in X : f_1(x) > \alpha\} = \{x \in X : f(x) > \alpha\} \setminus A.$$

If $\alpha < 0$, then

$$\{x \in X : f_1(x) > \alpha\} = \{x \in X : f(x) > \alpha\} \cup B.$$

Hence f_1 is measurable.

Conversely, if $A, B \in X$ and f_1 is measurable, then

$$\{x \in X : f(x) > \alpha\} = \{x \in X : f_1(x) > \alpha\} \cup A$$

when $\alpha \geqslant 0$, and

$$\{x \in X : f(x) > \alpha\} = \{x \in X : f_1(x) > \alpha\} \setminus B$$

when $\alpha < 0$.   Therefore f is measurable.   Q.E.D.

It is a consequence of Lemmas 2.6 and 2.8 that if f is in $M(X, X)$, then the functions

$$cf, \quad f^2, \quad |f|, \quad f^+, \quad f^-$$

also belong to $M(X, X)$.

The only comment that need be made is that we adopt the convention that $0(\pm\infty) = 0$ so that cf vanishes identically when $c = 0$.   If f and g belong to $M(X, X)$, then the sum $f + g$ is not well-defined by the formula $(f + g)(x) = f(x) + g(x)$ on the sets

$$\begin{aligned} E_1 &= \{x \in X : f(x) = -\infty \quad \text{and} \quad g(x) = +\infty\}, \\ E_2 &= \{x \in X : f(x) = +\infty \quad \text{and} \quad g(x) = -\infty\}, \end{aligned}$$

both of which belong to X.   However, if we define $f + g$ to be zero on $E_1 \cup E_2$, the resulting function on X is measurable.   We shall return to the measurability of the product fg after the next result.

2.9 LEMMA.   *Let (f_n) be a sequence in $M(X, X)$ and define the functions*

$$f(x) = \inf f_n(x), \quad F(x) = \sup f_n(x),$$
$$f^*(x) = \lim \inf f_n(x), \quad F^*(x) = \lim \sup f_n(x).$$

Then f, F, f^*, and F^* belong to $M(X, X)$.

PROOF.   Observe that

$$\{x \in X : f(x) \geqslant \alpha\} = \bigcap_{n=1}^{\infty} \{x \in X : f_n(x) \geqslant \alpha\},$$

$$\{x \in X : F(x) > \alpha\} = \bigcup_{n=1}^{\infty} \{x \in X : f_n(x) > \alpha\},$$

so that f and F are measurable when all the f_n are.   Since

$$f^*(x) = \sup_{n \geqslant 1} \left\{ \inf_{m \geqslant n} f_m(x) \right\},$$

$$F^*(x) = \inf_{n \geqslant 1} \left\{ \sup_{m \geqslant n} f_m(x) \right\},$$

the measurability of f^* and F^* is also established.   Q.E.D.

2.10 COROLLARY.   *If (f_n) is a sequence in $M(X, X)$ which converges to f on X, then f is in $M(X, X)$.*

PROOF.   In this case $f(x) = \lim f_n(x) = \lim \inf f_n(x)$.   Q.E.D.

We now return to the measurability of the product $f\,g$ when f, g belong to $M(X, X)$.   If $n \in N$, let f_n be the "truncation of f" defined by

$$f_n(x) = f(x), \quad \text{if} \ \ |f(x)| \leqslant n,$$
$$= n, \quad \text{if} \ \ f(x) > n,$$
$$= -n, \quad \text{if} \ \ f(x) < -n.$$

Let g_m be defined similarly.   It is readily seen that f_n and g_m are measurable (see Exercise 2.K).   It follows from Lemma 2.6 that the product $f_n g_m$ is measurable.   Since

$$f(x) g_m(x) = \lim_n f_n(x) g_m(x), \quad x \in X,$$

it follows from Corollary 2.10 that $f g_m$ belongs to $M(X, X)$. Since

$$(f g)(x) = f(x) g(x) = \lim_m f(x) g_m(x), \quad x \in X,$$

another application of Corollary 2.10 shows that $f g$ belongs to $M(X, X)$.

It has been seen that the limit of a sequence of functions in $M(X, X)$ belongs to $M(X, X)$. We shall now prove that a nonnegative function f in $M(X, X)$ is the limit of a monotone increasing sequence (φ_n) in $M(X, X)$. Moreover, each φ_n can be chosen to be nonnegative and to assume only a finite number of real values.

2.11 LEMMA. *If f is a nonnegative function in $M(X, X)$, then there exists a sequence (φ_n) in $M(X, X)$ such that*

(a) $0 \leqslant \varphi_n(x) \leqslant \varphi_{n+1}(x)$ *for $x \in X$, $n \in N$.*
(b) $f(x) = \lim \varphi_n(x)$ *for each $x \in X$.*
(c) *Each φ_n has only a finite number of real values.*

PROOF. Let n be a fixed natural number. If $k = 0, 1, \ldots, n2^n - 1$, let E_{kn} be the set

$$E_{kn} = \{x \in X : k2^{-n} \leqslant f(x) < (k + 1)2^{-n}\},$$

and if $k = n2^n$, let E_{kn} be the set $\{x \in X : f(x) \geqslant n\}$. We observe that the sets $\{E_{kn} : k = 0, 1, \ldots, n2^n\}$ are disjoint, belong to X, and have union equal to X. If we define φ_n to be equal to $k2^{-n}$ on E_{kn}, then φ_n belongs to $M(X, X)$. It is readily established that the properties (a), (b), (c) hold.   Q.E.D.

COMPLEX-VALUED FUNCTIONS

It is frequently important to consider complex-valued functions defined on X and to have a notion of measurability for such functions. We observe that if f is a complex-valued function defined on X, then there exist two uniquely determined real-valued functions f_1, f_2 such that

$$f = f_1 + i f_2.$$

(Indeed, $f_1(x) = \operatorname{Re} f(x)$, $f_2(x) = \operatorname{Im} f(x)$, for $x \in X$.) We define the complex-valued function f to be **measurable** if and only if its **real** and **imaginary parts** f_1 and f_2, respectively, are measurable. It is easy to see that sums, products, and limits of complex-valued measurable functions are also measurable.

FUNCTIONS BETWEEN MEASURABLE SPACES

In the sequel we shall require the notion of measurability only for real- and complex-valued functions. In some work, however, one wishes to define measurability for a function f from one measurable space (X, X) into another measurable space (Y, Y). In this case one says that f is **measurable** in case the set

$$f^{-1}(E) = \{x \in X : f(x) \in E\}$$

belongs to X for every set E belonging to Y. Although this definition of measurability appears to differ from Definition 2.3, it is not difficult to show (see Exercise 2.P) that Definition 2.3 is equivalent to this definition in the case that $Y = R$ and $Y = B$.

This definition of measurability shows very clearly the close analogy between the measurable functions on a measurable space and continuous functions on a topological space.

EXERCISES

2.A. Show that $[a, b] = \bigcap_{n=1}^{\infty} (a - 1/n, b + 1/n)$. Hence any σ-algebra of subsets of R which contains all open intervals also contains all closed intervals. Similarly, $(a, b) = \bigcup_{n=1}^{\infty} [a + 1/n, b - 1/n]$, so that any σ-algebra containing all closed intervals also contains all open intervals.

2.B. Show that the Borel algebra B is also generated by the collection of all half-open intervals $(a, b] = \{x \in R : a < x \leqslant b\}$. Also show that B is generated by the collection of all half-rays $\{x \in R : x > a\}$, $a \in R$.

2.C. Let (A_n) be a sequence of subsets of a set X. Let $E_0 = \emptyset$ and for $n \in N$, let

$$E_n = \bigcup_{k=1}^{n} A_k, \quad F_n = A_n \setminus E_{n-1}.$$

Show that (E_n) is a monotone increasing sequence of sets and that (F_n) is a disjoint sequence of sets (that is, $F_n \cap F_m = \emptyset$ if $n \neq m$) such that

$$\bigcup_{n=1}^{\infty} E_n = \bigcup_{n=1}^{\infty} F_n = \bigcup_{n=1}^{\infty} A_n.$$

2.D. Let (A_n) be a sequence of subsets of a set X. If A consists of all $x \in X$ which belong to infinitely many of the sets A_n, show that

$$A = \bigcap_{m=1}^{\infty} \left[\bigcup_{n=m}^{\infty} A_n \right].$$

The set A is often called the **limit superior** of the sets (A_n) and denoted by $\lim \sup A_n$.

2.E. Let (A_n) be a sequence of subsets of a set X. If B consists of all $x \in X$ which belong to all but a finite number of the sets A_n, show that

$$B = \bigcup_{m=1}^{\infty} \left[\bigcap_{n=m}^{\infty} A_n \right].$$

The set B is often called the **limit inferior** of the sets (A_n) and denoted by $\lim \inf A_n$.

2.F. If (E_n) is a sequence of subsets of a set X which is monotone increasing (that is, $E_1 \subseteq E_2 \subseteq E_3 \subseteq \cdots$), show that

$$\lim \sup E_n = \bigcup_{n=1}^{\infty} E_n = \lim \inf E_n.$$

2.G. If (F_n) is a sequence of subsets of a set X which is monotone decreasing (that is, $F_1 \supseteq F_2 \supseteq F_3 \supseteq \cdots$), show that

$$\lim \sup F_n = \bigcap_{n=1}^{\infty} F_n = \lim \inf F_n.$$

2.H. If (A_n) is a sequence of subsets of X, show that

$$\emptyset \subseteq \lim \inf A_n \subseteq \lim \sup A_n \subseteq X.$$

Give an example of a sequence (A_n) such that

$$\lim \inf A_n = \emptyset, \qquad \lim \sup A_n = X.$$

Give an example of a sequence (A_n) which is neither monotone increasing or decreasing, but is such that

$$\lim \inf A_n = \lim \sup A_n.$$

When this equality holds, the common value is called the **limit** of (A_n) and is denoted by $\lim A_n$.

2.I. Give an example of a function f on X to $\boldsymbol{R}$ which is not X-measurable, but is such that the functions $|f|$ and f^2 are X-measurable.

2.J. If a, b, c are real numbers, let mid (a, b, c) denote the "value in the middle." Show that

$$\text{mid } (a, b, c) = \inf \left\{\sup \{a, b\}, \sup \{a, c\}, \sup \{b, c\}\right\}.$$

If f_1, f_2, f_3 are X-measurable functions on X to $\boldsymbol{R}$ and if g is defined for $x \in X$ by

$$g(x) = \text{mid } \left(f_1(x), \quad f_2(x), \quad f_3(x)\right),$$

then g is X-measurable.

2.K. Show directly (without using the preceding exercise) that if f is measurable and $A > 0$, then the **truncation** f_A defined by

$$
\begin{aligned}
f_A(x) &= f(x), && \text{if } |f(x)| \leqslant A, \\
&= A, && \text{if } f(x) > A, \\
&= -A, && \text{if } f(x) < -A,
\end{aligned}
$$

is measurable.

2.L. Let f be a nonnegative X-measurable function on X which is bounded (that is, there exists a constant K such that $0 \leqslant f(x) \leqslant K$ for all x in X). Show that the sequence (φ_n) constructed in Lemma 2.11 converges *uniformly* on X to f.

2.M. Let f be a function defined on a set X with values in a set Y. If E is any subset of Y, let

$$f^{-1}(E) = \{x \in X : f(x) \in E\}.$$

Show that $f^{-1}(\emptyset) = \emptyset$, $f^{-1}(Y) = X$. If E and F are subsets of Y, then

$$f^{-1}(E \setminus F) = f^{-1}(E) \setminus f^{-1}(F).$$

If $\{E_\alpha\}$ is any nonempty collection of subsets of Y, then

$$f^{-1}\left(\bigcup_\alpha E_\alpha\right) = \bigcup_\alpha f^{-1}(E_\alpha), \quad f^{-1}\left(\bigcap_\alpha E_\alpha\right) = \bigcap_\alpha f^{-1}(E_\alpha).$$

In particular it follows that if Y is a σ-algebra of subsets of Y, then $\{f^{-1}(E) : E \in Y\}$ is a σ-algebra of subsets of X.

2.N. Let f be a function defined on a set X with values in a set Y. Let $\mathbf{X}$ be a σ-algebra of subsets of X and let $\mathbf{Y} = \{E \subseteq Y : f^{-1}(E) \in \mathbf{X}\}$. Show that $\mathbf{Y}$ is a σ-algebra.

2.O. Let $(X, \mathbf{X})$ be a measurable space and f be defined on X to Y. Let A be a collection of subsets of Y such that $f^{-1}(E) \in \mathbf{X}$ for every $E \in A$. Show that $f^{-1}(F) \in \mathbf{X}$ for any set F which belongs to the σ-algebra generated by A. (*Hint:* Use the preceding exercise.)

2.P. Let $(X, \mathbf{X})$ be a measurable space and f be a real-valued function defined on X. Show that f is $\mathbf{X}$-measurable if and only if $f^{-1}(E) \in \mathbf{X}$ for every Borel set E.

2.Q. Let $(X, \mathbf{X})$ be a measurable space, f be an $\mathbf{X}$-measurable function on X to $\mathbf{R}$ and let φ be a continuous function on $\mathbf{R}$ to $\mathbf{R}$. Show that the composition $\varphi \circ f$, defined by $(\varphi \circ f)(x) = \varphi[f(x)]$, is $\mathbf{X}$-measurable. (*Hint:* If φ is continuous, then $\varphi^{-1}(E) \in \mathbf{B}$ for each $E \in \mathbf{B}$.)

2.R. Let f be as in the preceding exercise and let ψ be a Borel measurable function. Show that $\psi \circ f$ is $\mathbf{X}$-measurable.

2.S. Let f be a complex-valued function defined on a measurable space $(X, \mathbf{X})$. Show that f is $\mathbf{X}$-measurable if and only if

$$\{x \in X : a < \operatorname{Re} f(x) < b, \quad c < \operatorname{Im} f(x) < d\}$$

belongs to $\mathbf{X}$ for all real numbers a, b, c, d. More generally, f is $\mathbf{X}$-measurable if and only if $f^{-1}(G) \in \mathbf{X}$ for every open set G in the complex plane $\mathbf{C}$.

2.T. Show that sums, products, and limits of complex-valued measurable functions are measurable.

2.U. Show that a function f on X to $\mathbf{R}$ (or to $\bar{\mathbf{R}}$) is $\mathbf{X}$-measurable if and only if the set A_α in Lemma 2.4(a) belongs to $\mathbf{X}$ for each rational number α; or, if and only if the set B_α in Lemma 2.4(b) belongs to $\mathbf{X}$ for each rational number α; etc.

2.V. A nonempty collection $\mathbf{M}$ of subsets of a set X is called a **monotone class** if, for each monotone increasing sequence (E_n) in $\mathbf{M}$ and each monotone decreasing sequence (F_n) in $\mathbf{M}$, the sets

$$\bigcup_{n=1}^{\infty} E_n, \qquad \bigcap_{n=1}^{\infty} F_n$$

belong to M. Show that a σ-algebra is a monotone class. Also, if A is a nonempty collection of subsets of X, then there is a smallest monotone class containing A. (This smallest monotone class is called the **monotone class generated by** A.)

2.W. If A is a nonempty collection of subsets of X, then the σ-algebra S generated by A contains the monotone class M generated by A. Show that the inclusion $A \subseteq M \subseteq S$ may be proper.

CHAPTER 3

Measures

We have introduced the notion of a measurable space (X, X) consisting of a set X and a σ-algebra X of subsets of X. We now consider certain functions which are defined on X and have real, or extended real values. These functions, which will be called "measures," are suggested by our idea of length, area, mass, and so forth. Thus it is natural that they should attach the value 0 to the empty set $\emptyset$ and that they should be additive over disjoint sets in X. (Actually we shall require that they be countably additive in the sense to be described below.) It is also desirable to permit the measures to take on the extended real number $+\infty$.

3.1 DEFINITION. A **measure** is an extended real-valued function μ defined on a σ-algebra X of subsets of X such that (i) $\mu(\emptyset) = 0$, (ii) $\mu(E) \geq 0$ for all $E \in X$, and (iii) μ is **countably additive** in the sense that if (E_n) is any disjoint sequence* of sets in X, then

$$(3.1) \qquad \mu\left(\bigcup_{n=1}^{\infty} E_n \right) = \sum_{n=1}^{\infty} \mu(E_n).$$

Since we permit μ to take on $+\infty$, we remark that the appearance of the value $+\infty$ on the right side of the equation (3.1) means either that $\mu(E_n) = +\infty$ for some n or that the series of nonnegative terms on the right side of (3.1) is divergent. If a measure does not take on $+\infty$,

---

* This means that $E_n \cap E_m = \emptyset$ if $n \neq m$.

we say that it is **finite**.   More generally, if there exists a sequence (E_n) of sets in X with $X = \bigcup E_n$ and such that $\mu(E_n) < +\infty$ for all n, then we say that μ is σ-**finite**.

3.2 EXAMPLES.   (a) Let X be any nonempty set and let X be the σ-algebra of all subsets of X.   Let μ_1 be defined on X by

$$\mu_1(E) = 0, \qquad \text{for all } E \in X;$$

and let μ_2 be defined by

$$\mu_2(\emptyset) = 0, \quad \mu_2(E) = +\infty \qquad \text{if} \quad E \neq \emptyset.$$

Both μ_1 and μ_2 are measures, although neither one is very interesting. Note that μ_2 is neither finite nor σ-finite.

(b) Let (X, X) be as in (a) and let p be a fixed element of X.   Let μ be defined for $E \in X$ by

$$\mu(E) = 0, \qquad \text{if} \quad p \notin E,$$
$$= 1, \qquad \text{if} \quad p \in E.$$

It is readily seen that μ is a finite measure; it is called the **unit measure concentrated at** p.

(c) Let $X = N = \{1, 2, 3, \dots\}$ and let X be the σ-algebra of all subsets of N.   If $E \in X$, define $\mu(E)$ to be equal to the number of elements in E if E is a finite set and equal to $+\infty$ if E is an infinite set. Then μ is a measure and is called the **counting measure on** N.   Note that μ is not finite, but it is σ-finite.

(d) If $X = R$ and $X = B$, the Borel algebra, then it will be shown in Chapter 9 that there exists a unique measure λ defined on B which coincides with length on open intervals.   [By this we mean that if E is the nonempty interval (a, b), then $\lambda(E) = b - a$.]   This unique measure is usually called **Lebesgue** (or **Borel**) **measure**.   It is not a finite measure, but it is σ-finite.

(e) If $X = R$, $X = B$, and f is a continuous monotone increasing function, then it will be shown in Chapter 9 that there exists a unique measure λ_f defined on B such that if $E = (a, b)$, then $\lambda_f(E) = f(b) - f(a)$.   This measure λ_f is called the **Borel-Stieltjes measure generated by** f.

We shall now derive a few simple results that will be needed later.

3.3 LEMMA. *Let μ be a measure defined on a σ-algebra X. If E and F belong to X and $E \subseteq F$, then $\mu(E) \leqslant \mu(F)$. If $\mu(E) < +\infty$, then $\mu(F \setminus E) = \mu(F) - \mu(E)$.*

PROOF. Since $F = E \cup (F \setminus E)$ and $E \cap (F \setminus E) = \emptyset$, it follows that

$$\mu(F) = \mu(E) + \mu(F \setminus E).$$

Since $\mu(F \setminus E) \geqslant 0$, it follows that $\mu(F) \geqslant \mu(E)$. If $\mu(E) < +\infty$, then we can subtract it from both sides of this equation.   Q.E.D.

3.4 LEMMA. *Let μ be a measure defined on a σ-algebra X.*

(a) *If (E_n) is an increasing sequence in X, then*

(3.2)
$$\mu\left(\bigcup_{n=1}^{\infty} E_n \right) = \lim \mu(E_n).$$

(b) *If (F_n) is a decreasing sequence in X and if $\mu(F_1) < +\infty$, then*

(3.3)
$$\mu\left(\bigcap_{n=1}^{\infty} F_n \right) = \lim \mu(F_n).$$

PROOF. (a) If $\mu(E_n) = +\infty$ for some n, then both sides of equation (3.2) are $+\infty$. Hence we can suppose that $\mu(E_n) < +\infty$ for all n.

Let $A_1 = E_1$ and $A_n = E_n \setminus E_{n-1}$ for $n > 1$. Then (A_n) is a disjoint sequence of sets in X such that

$$E_n = \bigcup_{j=1}^{n} A_j, \qquad \bigcup_{n=1}^{\infty} E_n = \bigcup_{n=1}^{\infty} A_n.$$

Since μ is countably additive,

$$\mu\left(\bigcup_{n=1}^{\infty} E_n \right) = \sum_{n=1}^{\infty} \mu(A_n) = \lim \sum_{n=1}^{m} \mu(A_n).$$

By Lemma 3.3 $\mu(A_n) = \mu(E_n) - \mu(E_{n-1})$ for $n > 1$, so the finite series on the right side telescopes and

$$\sum_{n=1}^{m} \mu(A_n) = \mu(E_m).$$

Hence equation (3.2) is proved.

(b) Let $E_n = F_1 \setminus F_n$, so that (E_n) is an increasing sequence of sets in X. If we apply part (a) and Lemma 3.3, we infer that

$$\mu\left(\bigcup_{n=1}^{\infty} E_n \right) = \lim \mu(E_n) = \lim [\mu(F_1) - \mu(F_n)]$$
$$= \mu(F_1) - \lim \mu(F_n).$$

Since $\bigcup_{n=1}^{\infty} E_n = F_1 \setminus \bigcap_{n=1}^{\infty} F_n$, it follows that

$$\mu\left(\bigcup_{n=1}^{\infty} E_n \right) = \mu(F_1) - \mu\left(\bigcap_{n=1}^{\infty} F_n \right).$$

Combining these two equations, we obtain (3.3).   Q.E.D.

3.5 DEFINITION. A **measure space** is a triple (X, X, μ) consisting of a set X, a σ-algebra X of subsets of X, and a measure μ defined on X.

There is a terminological matter that needs to be mentioned and which shall be employed in the following. We shall say that a certain proposition holds μ-**almost everywhere** if there exists a subset $N \in X$ with $\mu(N) = 0$ such that the proposition holds on the complement of N. Thus we say that two functions f, g are **equal** μ-**almost everywhere** or that they are **equal for** μ-**almost all** x in case $f(x) = g(x)$ when $x \notin N$, for some $N \in X$ with $\mu(N) = 0$. In this case we will often write

$$f = g, \quad \mu\text{-a.e.}$$

In like manner, we say that a sequence (f_n) of functions on X **converges** μ-**almost everywhere** (or **converges for** μ-**almost all** x) if there exists a set $N \in X$ with $\mu(N) = 0$ such that $f(x) = \lim f_n(x)$ for $x \notin N$. In this case we often write

$$f = \lim f_n, \quad \mu\text{-a.e.}$$

Of course, if the measure μ is understood, we shall say "almost everywhere" instead of "μ-almost everywhere."

There are some instances (suggested by the notion of electrical charge, for example) in which it is desirable to discuss functions which behave like measures except that they take both positive and negative values. In this case, it is not so convenient to permit the extended real numbers $+\infty$, $-\infty$ to be values since we wish to avoid expressions of the form $(+\infty) + (-\infty)$. Although it is possible to handle "signed measures" which take on only *one* of the values $+\infty$, $-\infty$, we shall restrict our attention to the case where neither of these symbols is permitted. To indicate this restriction, we shall introduce the term "charge," which is not entirely standard.

3.6 Definition. If X is a σ-algebra of subsets of a set X, then a real-valued function λ defined on X is said to be a **charge** in case $\lambda(\emptyset) = 0$ and λ is countably additive in the sense that if (E_n) is a disjoint sequence of sets in X, then

$$\lambda\left(\bigcup_{n=1}^{\infty} E_n \right) = \sum_{n=1}^{\infty} \lambda(E_n).$$

[Since the left-hand side is independent of the order and this equality is required for all such sequences, the series on the right-hand side must be unconditionally convergent for all disjoint sequences of measurable sets.]

It is clear that the sum and difference of two charges is a charge. More generally, any finite linear combination of charges is a charge. It will be seen in Chapter 5 that functions which are integrable over a measure space (X, X, μ) give rise to charges. Later, in Chapter 8, we will characterize those charges which are generated by integrable functions.

EXERCISES

3.A. If μ is a measure on X and A is a fixed set in X, then the function λ, defined for $E \in X$ by $\lambda(E) = \mu(A \cap E)$, is a measure on X.

3.B. If $\mu_1, \ldots, \mu_n$ are measures on X and $a_1, \ldots, a_n$ are nonnegative real numbers, then the function λ, defined for $E \in X$ by

$$\lambda(E) = \sum_{j=1}^{n} a_j \mu_j(E),$$

is a measure on X.

3.C. If (μ_n) is a sequence of measures on X with $\mu_n(X) = 1$ and if λ is defined by

$$\lambda(E) = \sum_{n=1}^{\infty} 2^{-n} \mu_n(E), \quad E \in X,$$

then λ is a measure on X and $\lambda(X) = 1$.

3.D. Let $X = N$ and let X be the σ-algebra of all subsets of N. If (a_n) is a sequence of nonnegative real numbers and if we define μ by

$$\mu(\emptyset) = 0; \quad \mu(E) = \sum_{n \in E} a_n, \quad E \neq \emptyset;$$

then μ is a measure on X. Conversely, every measure on X is obtained in this way for some sequence (a_n) in $\overline{R}^+$.

3.E. Let X be an uncountable set and let X be the family of all subsets of X. Define μ on E in X by requiring that $\mu(E) = 0$, if E is countable, and $\mu(E) = +\infty$, if E is uncountable. Show that μ is a measure on X.

3.F. Let $X = N$ and let X be the family of all subsets of N. If E is finite, let $\mu(E) = 0$; if E is infinite, let $\mu(E) = +\infty$. Is μ a measure on X?

3.G. If X and X are as in Exercise 3.F, let $\lambda(E) = +\infty$ for all $E \in X$. Is λ a measure?

3.H. Show that Lemma 3.4(b) may fail if the finiteness condition $\mu(F_1) < +\infty$ is dropped.

3.I. Let (X, X, μ) be a measure space and let (E_n) be a sequence in X. Show that
$$\mu(\liminf E_n) \leqslant \liminf \mu(E_n).$$
[See Exercise 2.E.]

3.J. Using the notation of Exercise 2.D, show that
$$\limsup \mu(E_n) \leqslant \mu(\limsup E_n)$$
when $\mu(\bigcup E_n) < +\infty$. Show that this inequality may fail if $\mu(\bigcup E_n) = +\infty$.

3.K. Let (X, X, μ) be a measure space and let $Z = \{E \in X : \mu(E) = 0\}$. Is Z a σ-algebra? Show that if $E \in Z$ and $F \in X$, then $E \cap F \in Z$. Also, if E_n belongs to Z for $n \in N$, then $\bigcup E_n \in Z$.

3.L. Let X, X, μ, Z be as in Exercise 3.K and let X' be the family of all subsets of X of the form
$$(E \cup Z_1) \setminus Z_2, \quad E \in X,$$

where Z_1 and Z_2 are arbitrary subsets of sets belonging to Z. Show that a set is in X' if and only if it has the form $E \cup Z$ where $E \in X$ and Z is a subset of a set in Z. Show that the collection X' forms a σ-algebra of sets in X. The σ-algebra X' is called the **completion** of X (with respect to μ).

3.M. With the notation of Exercise 3.L, let μ' be defined on X' by

$$\mu'(E \cup Z) = \mu(E),$$

when $E \in X$ and Z is a subset of a set in Z. Show that μ' is well-defined and is a measure on X' which agrees with μ on X. The measure μ' is called the **completion** of μ.

3.N. Let (X, X, μ) be a measure space and let (X, X', μ') be its completion in the sense of Exercise 3.M. Suppose that f is an X'-measurable function on X to $\bar{R}$. Show that there exists an X-measurable function g on X to $\bar{R}$ which is μ-almost everywhere equal to f. (*Hint:* For each rational number r, let $A_r = \{x : f(x) > r\}$ and write $A_r = E_r \cup Z_r$, where $E_r \in X$ and Z_r is a subset of a set in Z. Let Z be a set in Z containing $\bigcup Z_r$ and define $g(x) = f(x)$ for $x \notin Z$, and $g(x) = 0$ for $x \in Z$. To show that g is X-measurable, use Exercise 2.U.)

3.O. Show that Lemma 3.4 holds if μ is a charge on X.

3.P. If μ is a charge on X, let π be defined for $E \in X$ by

$$\pi(E) = \sup \{\mu(A) : A \subseteq E, A \in X\}.$$

Show that π is a measure on X. (*Hint:* If $\pi(E_n) < \infty$ and $\varepsilon > 0$, let $F_n \in X$ be such that $F_n \subseteq E_n$ and $\pi(E_n) \leqslant \mu(F_n) + 2^{-n}\varepsilon$.)

3.Q. If μ is a charge on X, let ν be defined for $E \in X$ by

$$\nu(E) = \sup \sum_{j=1}^{n} |\mu(A_j)|,$$

where the supremum is taken over all finite disjoint collections $\{A_j\}$ in X with $E = \bigcup_{j=1}^{n} A_j$. Show that ν is a measure on X. (It is called the **variation of** μ.)

3.R. Let λ denote Lebesgue measure defined on the Borel algebra B of R [see Example 3.2(d)]. (a) If E consists of a single point, then $E \in B$ and $\lambda(E) = 0$. (b) If E is countable, then $E \in B$ and $\lambda(E) = 0$.

(c) The open interval (a, b), the half-open intervals $(a, b]$, $[a, b)$, and the closed interval $[a, b]$ all have Lebesgue measure $b - a$.

3.S. If λ denotes Lebesgue measure and E is an open subset of $\boldsymbol{R}$, then $\lambda(E) > 0$ if and only if E is nonvoid.   Show that if K is a compact subset of $\boldsymbol{R}$, then $\lambda(K) < +\infty$.

3.T. Show that the Lebesgue measure of the Cantor set (see Reference [1], p. 52) is zero.

3.U. By varying the construction of the Cantor set, obtain a set of positive Lebesgue measure which contains no nonvoid open interval.

3.V. Suppose that E is a subset of a set $N \in X$ with $\mu(N) = 0$ but that $E \notin X$.   The sequence $(f_n), f_n = 0$, converges μ-almost everywhere to χ_E.   Hence the almost everywhere limit of a sequence of measurable functions may not be measurable.

CHAPTER 4

The Integral

In this chapter we shall introduce the integral first for nonnegative simple measurable functions and then for arbitrary nonnegative extended real-valued measurable functions. The principal result is the celebrated Monotone Convergence Theorem, which is a basic tool for everything that follows.

Throughout this chapter we shall consider a fixed measure space $(X, \mathbf{X}, \mu)$. We shall denote the collection of all $\mathbf{X}$-measurable functions on X to $\bar{\mathbf{R}}$ by $M = M(X, \mathbf{X})$ and the collection of all nonnegative $\mathbf{X}$-measurable functions on X to $\bar{\mathbf{R}}$ by $M^+ = M^+(X, \mathbf{X})$. We shall define the integral of any function in M^+ with respect to the measure μ. In order to do so we shall find it convenient to introduce the notion of a simple function. It is convenient to require that simple functions have values in $\mathbf{R}$ rather than in $\bar{\mathbf{R}}$.

4.1 DEFINITION. A real-valued function is **simple** if it has only a finite number of values.

A simple measurable function φ can be represented in the form

$$(4.1) \qquad\qquad \varphi = \sum_{j=1}^{n} a_j \chi_{E_j},$$

where $a_j \in \mathbf{R}$ and χ_{E_j} is the characteristic function of a set E_j in $\mathbf{X}$. Among these representations for φ there is a unique **standard representation** characterized by the fact that the a_j are distinct and the E_j

are disjoint nonempty subsets of X and are such that $X = \bigcup_{j=1}^{n} E_j$. (Of course, if we do not require the a_j to be distinct, or the sets E_j to be disjoint, then a simple function has many representations as a linear combination of characteristic functions.)

4.2 DEFINITION.   If φ is a simple function in $M^{+}(X, X)$ with the standard representation (4.1), we define the **integral** of φ with respect to μ to be the extended real number

$$(4.2) \qquad \int \varphi \, d\mu = \sum_{j=1}^{n} a_j \, \mu(E_j).$$

In the expression (4.2) we employ the convention that $0(+\infty) = 0$ so the integral of the function identically 0 is equal to 0 whether the space has finite or infinite measure.   It should be noted that the value of the integral of a simple function in M^{+} is well-defined (although it may be $+\infty$) since all the a_j are nonnegative, and so we do not encounter meaningless expressions such as $(+\infty) - (+\infty)$.

We shall need the following elementary properties of the integral.

4.3 LEMMA.   (a) *If φ and ψ are simple functions in $M^{+}(X, X)$ and $c \geqslant 0$, then*

$$\int c\varphi \, d\mu = c \int \varphi \, d\mu,$$

$$\int (\varphi + \psi) \, d\mu = \int \varphi \, d\mu + \int \psi \, d\mu.$$

(b) *If λ is defined for E in X by*

$$\lambda(E) = \int \varphi \, \chi_E \, d\mu,$$

then λ is a measure on X.

PROOF.   If $c = 0$, then $c\varphi$ vanishes identically and the equality holds.   If $c > 0$, then $c\varphi$ is in M^{+} with standard representation

$$c\varphi = \sum_{j=1}^{n} ca_j \, \chi_{E_j},$$

when φ has standard representation (4.1). Therefore

$$\int c\varphi \, d\mu = \sum_{j=1}^{n} ca_j \, \mu(E_j) = c \sum_{j=1}^{n} a_j \, \mu(E_j) = c \int \varphi \, d\mu.$$

Let φ and ψ have standard representations

$$\varphi = \sum_{j=1}^{n} a_j \, \chi_{E_j}, \qquad \psi = \sum_{k=1}^{m} b_k \, \chi_{F_k},$$

then $\varphi + \psi$ has a representation

$$\varphi + \psi = \sum_{j=1}^{n} \sum_{k=1}^{m} (a_j + b_k)\chi_{E_j \cap F_k}.$$

However, this representation of $\varphi + \psi$ as a linear combination of characteristic functions of the disjoint sets $E_j \cap F_k$ is not necessarily the standard representation for $\varphi + \psi$, since the values $a_j + b_k$ may not be distinct. Let c_h, $h = 1, \ldots, p$, be the distinct numbers in the set $\{a_j + b_k : j = 1, \ldots, n; k = 1, \ldots, m\}$ and let G_h be the union of all those sets $E_j \cap F_k \neq \emptyset$ such that $a_j + b_k = c_h$. Thus

$$\mu(G_h) = \sum_{(h)} \mu(E_j \cap F_k),$$

where the notation designates summation over all j, k such that $a_j + b_k = c_h$. Since the standard representation of $\varphi + \psi$ is given by

$$\varphi + \psi = \sum_{h=1}^{p} c_h \, \chi_{G_h},$$

we find that

$$\begin{aligned}
\int (\varphi + \psi) \, d\mu &= \sum_{h=1}^{p} c_h \, \mu(G_h) = \sum_{h=1}^{p} \sum_{(h)} c_h \, \mu(E_j \cap F_k) \\
&= \sum_{h=1}^{p} \sum_{(h)} (a_j + b_k) \, \mu(E_j \cap F_k) \\
&= \sum_{j=1}^{n} \sum_{k=1}^{m} (a_j + b_k) \, \mu(E_j \cap F_k) \\
&= \sum_{j=1}^{n} \sum_{k=1}^{m} a_j \, \mu(E_j \cap F_k) + \sum_{j=1}^{n} \sum_{k=1}^{m} b_k \, \mu(E_j \cap F_k).
\end{aligned}$$

Since X is the union of both of the disjoint families $\{E_j\}$ and $\{F_k\}$, then

$$\mu(E_j) = \sum_{k=1}^{m} \mu(E_j \cap F_k), \quad \mu(F_k) = \sum_{j=1}^{n} \mu(E_j \cap F_k).$$

We employ this observation (and change the order of summation in the second term) to obtain the desired relation

$$\int (\varphi + \psi) \, d\mu = \sum_{j=1}^{n} a_j \, \mu(E_j) + \sum_{k=1}^{m} b_k \, \mu(F_k)$$

$$= \int \varphi \, d\mu + \int \psi \, d\mu.$$

To establish part (b), we observe that

$$\varphi \, \chi_E = \sum_{j=1}^{n} a_j \, \chi_{E_j \cap E}.$$

Hence, it follows by induction from what we have proved that

$$\lambda(E) = \int \varphi \, \chi_E \, d\mu = \sum_{j=1}^{n} a_j \int \chi_{E_j \cap E} \, d\mu = \sum_{j=1}^{n} a_j \, \mu(E_j \cap E).$$

Since the mapping $E \to \mu(E_j \cap E)$ is a measure (see Exercise 3.A) we have expressed λ as a linear combination of measures on X. It follows (see Exercise 3.B) that λ is also a measure on X.   Q.E.D.

We are now prepared to introduce the integral of an arbitrary function in M^+. Observe that we do not require the value of the integral to be finite.

4.4 DEFINITION. If f belongs to $M^+(X, X)$, we define the **integral of f with respect to** μ to be the extended real number

(4.3) $$\int f \, d\mu = \sup \int \varphi \, d\mu,$$

where the supremum is extended over all simple functions φ in $M^+(X, X)$ satisfying $0 \leqslant \varphi(x) \leqslant f(x)$ for all $x \in X$. If f belongs to $M^+(X, X)$ and E belongs to X, then $f\chi_E$ belongs to $M^+(X, X)$ and we define the **integral of f over E with respect to** μ to be the extended real number

(4.4) $$\int_E f \, d\mu = \int f \chi_E \, d\mu.$$

We shall first show that the integral is monotone both with respect to the integrand and the set over which the integral is extended.

4.5 LEMMA. (a) *If f and g belong to $M^+(X, X)$ and $f \leqslant g$, then*

(4.5)
$$\int f \, d\mu \leqslant \int g \, d\mu.$$

(b) *If f belongs to $M^+(X, X)$, if E, F belong to X, and if $E \subseteq F$, then*

$$\int_E f \, d\mu \leqslant \int_F f \, d\mu.$$

PROOF. (a) If φ is a simple function in M^+ such that $0 \leqslant \varphi \leqslant f$, then $0 \leqslant \varphi \leqslant g$. Therefore (4.5) holds.

(b) Since $f\chi_E \leqslant f\chi_F$, part (b) follows from (a).   Q.E.D.

We are now prepared to establish an important result due to B. Levi. This theorem provides the key to the fundamental convergence properties of the Lebesgue integral.

4.6 MONOTONE CONVERGENCE THEOREM. *If (f_n) is a monotone increasing sequence of functions in $M^+(X, X)$ which converges to f, then*

(4.6)
$$\int f \, d\mu = \lim \int f_n \, d\mu.$$

PROOF. According to Corollary 2.10, the function f is measurable. Since $f_n \leqslant f_{n+1} \leqslant f$, it follows from Lemma 4.5(a) that

$$\int f_n \, d\mu \leqslant \int f_{n+1} \, d\mu \leqslant \int f \, d\mu$$

for all $n \in N$. Therefore we have

$$\lim \int f_n \, d\mu \leqslant \int f \, d\mu.$$

To establish the opposite inequality, let α be a real number satisfying $0 < \alpha < 1$ and let φ be a simple measurable function satisfying $0 \leqslant \varphi \leqslant f$. Let

$$A_n = \{x \in X : f_n(x) \geqslant \alpha\varphi(x)\}$$

so that $A_n \in X$, $A_n \subseteq A_{n+1}$, and $X = \bigcup A_n$.   According to Lemma 4.5,

(4.7) $$\int_{A_n} \alpha\varphi \, d\mu \leq \int_{A_n} f_n \, d\mu \leq \int f_n \, d\mu.$$

Since the sequence (A_n) is monotone increasing and has union X, it follows from Lemmas 4.3(b) and 3.4(a) that

$$\int \varphi \, d\mu = \lim \int_{A_n} \varphi \, d\mu.$$

Therefore, on taking the limit in (4.7) with respect to n, we obtain

$$\alpha \int \varphi \, d\mu \leq \lim \int f_n \, d\mu.$$

Since this holds for all α with $0 < \alpha < 1$, we infer that

$$\int \varphi \, d\mu \leq \lim \int f_n \, d\mu$$

and since φ is an arbitrary simple function in M^+ satisfying $0 \leq \varphi \leq f$, we conclude that

$$\int f \, d\mu = \sup_{\varphi} \int \varphi \, d\mu \leq \lim \int f_n \, d\mu.$$

If we combine this with the opposite inequality, we obtain (4.6).   Q.E.D.

REMARK.   It should be observed that it is not being assumed that either side of (4.6) is finite.   Indeed, the sequence $(\int f_n \, d\mu)$ is a monotone increasing sequence of extended real numbers and so always has a limit in $\bar{R}$, but perhaps not in R.

We shall now derive some consequences of the Monotone Convergence Theorem.

4.7 COROLLARY.   (a) *If f belongs to M^+ and $c \geq 0$, then cf belongs to M^+ and*

$$\int cf \, d\mu = c \int f \, d\mu.$$

(b) *If f, g belong to M^+, then $f + g$ belongs to M^+ and*

$$\int (f + g) \, d\mu = \int f \, d\mu + \int g \, d\mu.$$

PROOF. (a) If $c = 0$ the result is immediate. If $c > 0$, let (φ_n) be a monotone increasing sequence of simple functions in M^+ converging to f on X (see Lemma 2.11). Then $(c\varphi_n)$ is a monotone sequence converging to cf. If we apply Lemma 4.3(a) and the Monotone Convergence Theorem, we obtain

$$\int cf \, d\mu = \lim \int c \, \varphi_n \, d\mu$$
$$= c \lim \int \varphi_n \, d\mu = c \int f \, d\mu.$$

(b) If (φ_n) and (ψ_n) are monotone increasing sequences of simple functions converging to f and g, respectively, then $(\varphi_n + \psi_n)$ is a monotone increasing sequence converging to $f + g$. It follows from Lemma 4.3(a) and the Monotone Convergence Theorem that

$$\int (f + g) \, d\mu = \lim \int (\varphi_n + \psi_n) \, d\mu$$
$$= \lim \int \varphi_n \, d\mu + \lim \int \psi_n \, d\mu$$
$$= \int f \, d\mu + \int g \, d\mu. \qquad \text{Q.E.D.}$$

The next result, a consequence of the Monotone Convergence Theorem, is very important for it enables us to handle sequences of functions that are not monotone.

4.8 FATOU'S LEMMA. *If (f_n) belongs to $M^+(X, X)$, then*

(4.8) $$\int (\liminf f_n) \, d\mu \leqslant \liminf \int f_n \, d\mu.$$

PROOF. Let $g_m = \inf\{f_m, f_{m+1}, \ldots\}$ so that $g_m \leqslant f_n$ whenever $m \leqslant n$. Therefore

$$\int g_m \, d\mu \leqslant \int f_n \, d\mu, \qquad m \leqslant n,$$

so that

$$\int g_m \, d\mu \leqslant \liminf \int f_n \, d\mu.$$

Since the sequence (g_m) is increasing and converges to $\liminf f_n$, the

Monotone Convergence Theorem implies that

$$\int (\lim \inf f_n)\, d\mu = \lim \int g_m\, d\mu$$

$$\leqslant \lim \inf \int f_n\, d\mu. \qquad \text{Q.E.D.}$$

It will be seen in an exercise that Fatou's Lemma may fail if it is not assumed that $f_n \geqslant 0$.

4.9 COROLLARY. *If f belongs to M^+ and if λ is defined on X by*

$$(4.9) \qquad\qquad \lambda(E) = \int_E f\, d\mu,$$

then λ is a measure.

PROOF. Since $f \geqslant 0$ it follows that $\lambda(E) \geqslant 0$. If $E = \emptyset$, then $f\chi_E$ vanishes everywhere so that $\lambda(\emptyset) = 0$. To see that λ is countably additive, let (E_n) be a disjoint sequence of sets in X with union E and let f_n be defined to be

$$f_n = \sum_{k=1}^{n} f\chi_{E_k}.$$

It follows from Corollary 4.7(b) and induction that

$$\int f_n\, d\mu = \sum_{k=1}^{n} \int f\chi_{E_k}\, d\mu = \sum_{k=1}^{n} \lambda(E_k).$$

Since (f_n) is an increasing sequence in M^+ converging to $f\chi_E$, the Monotone Convergence Theorem implies that

$$\lambda(E) = \int f\chi_E\, d\mu = \lim \int f_n\, d\mu = \sum_{k=1}^{\infty} \lambda(E_k). \qquad \text{Q.E.D.}$$

4.10 COROLLARY. *Suppose that f belongs to M^+. Then $f(x) = 0$ μ-almost everywhere on X if and only if*

$$(4.10) \qquad\qquad \int f\, d\mu = 0.$$

PROOF. If equation (4.10) holds, let

$$E_n = \left\{ x \in X : f(x) > \frac{1}{n} \right\},$$

so that $f \geqslant (1/n) \chi_{E_n}$, from which

$$0 = \int f \, d\mu \geqslant \frac{1}{n} \mu(E_n) \geqslant 0.$$

It follows that $\mu(E_n) = 0$; hence the set

$$\{x \in X : f(x) > 0\} = \bigcup_{n=1}^{\infty} E_n$$

also has measure 0.

Conversely, let $f(x) = 0$ μ-almost everywhere. If

$$E = \{x \in X : f(x) > 0\},$$

then $\mu(E) = 0$. Let $f_n = n \chi_E$. Since $f \leqslant \lim \inf f_n$, it follows from Fatou's Lemma that

$$0 \leqslant \int f \, d\mu \leqslant \lim \inf \int f_n \, d\mu = 0. \qquad \text{Q.E.D.}$$

4.11 COROLLARY. *Suppose that f belongs to M^+, and define λ on X by equation (4.9). Then the measure λ is absolutely continuous with respect to μ in the sense that if $E \in X$ and $\mu(E) = 0$, then $\lambda(E) = 0$.*

PROOF. If $\mu(E) = 0$ for some $E \in X$, then $f\chi_E$ vanishes μ-almost everywhere. By Corollary 4.10, we have

$$\lambda(E) = \int f \chi_E \, d\mu = 0. \qquad \text{Q.E.D.}$$

We shall now show that the Monotone Convergence Theorem holds if convergence on X is replaced by almost everywhere convergence.

4.12 COROLLARY. *If (f_n) is a monotone increasing sequence of functions in $M^+(X, X)$ which converges μ-almost everywhere on X to a function f in M^+, then*

$$\int f \, d\mu = \lim \int f_n \, d\mu.$$

PROOF. Let $N \in X$ be such that $\mu(N) = 0$ and (f_n) converges to f at every point of $M = X \setminus N$. Then $(f_n \chi_M)$ converges to $f\chi_M$ on X, so the Monotone Convergence Theorem implies that

$$\int f \chi_M \, d\mu = \lim \int f_n \chi_M \, d\mu.$$

Since $\mu(N) = 0$, the functions $f\chi_N$ and $f_n\chi_N$ vanish μ-almost everywhere. It follows from Corollary 4.10 that

$$\int f\chi_N \, d\mu = 0, \qquad \int f_n \chi_N \, d\mu = 0.$$

Since $f = f\chi_M + f\chi_N$ and $f_n = f_n\chi_M + f_n\chi_N$, it follows that

$$\int f \, d\mu = \int f\chi_M \, d\mu = \lim \int f_n \chi_M \, d\mu = \lim \int f_n \, d\mu. \qquad \text{Q.E.D.}$$

4.13 COROLLARY. *Let (g_n) be a sequence in M^+, then*

$$\int \left(\sum_{n=1}^{\infty} g_n \right) d\mu = \sum_{n=1}^{\infty} \left(\int g_n \, d\mu \right).$$

PROOF. Let $f_n = g_1 + \cdots + g_n$, and apply the Monotone Convergence Theorem. Q.E.D.

EXERCISES

4.A. If the simple function φ in $M^+(X, X)$ has the (not necessarily standard) representation

$$\varphi = \sum_{k=1}^{m} b_k \chi_{F_k},$$

where $b_k \in R$ and $F_k \in X$, show that

$$\int \varphi \, d\mu = \sum_{k=1}^{m} b_k \, \mu(F_k).$$

4.B. The sum, scalar multiple, and product of simple functions are simple functions. [In other words, the simple functions in $M(X, X)$ form a vector subspace of $M(X, X)$, closed under products.]

4.C. If φ_1 and φ_2 are simple functions in $M(X, X)$, then

$$\psi = \sup \{\varphi_1, \varphi_2\}, \qquad \omega = \inf \{\varphi_1, \varphi_2\}$$

are also simple functions in $M(X, X)$.

4.D. If $f \in M^+$ and $c > 0$, then the mapping $\varphi \to \psi = c\varphi$ is a one-to-one mapping between simple functions $\varphi \in M^+$ with $\varphi \leqslant f$

and simple functions ψ in M^+ with $\psi \leqslant cf$. Use this observation to give a different proof of Corollary 4.7(a).

4.E. Let $f, g \in M^+$, let $\omega \in M^+$ be a simple function such that $\omega \leqslant f + g$ and let $\varphi_n(x) = \sup\{(m/n)\omega(x) : \text{for } 0 \leqslant m \leqslant n \text{ with } (m/n)\omega(x) \leqslant f(x)\}$. Also let $\psi_n(x) = \sup\{(1-1/n)\omega(x) - \varphi_n(x), 0\}$. Show that $(1 - 1/n)\omega \leqslant \varphi_n + \psi_n$ and $\varphi_n \leqslant f$, $\psi_n \leqslant g$.

4.F. Employ Exercise 4.E to establish Corollary 4.7(b) without using the Monotone Convergence Theorem.

4.G. Let $X = N$, let X be all subsets of N, and let μ be the counting measure on X. If f is a nonnegative function on N, then $f \in M^+(X, X)$ and

$$\int f \, d\mu = \sum_{n=1}^{\infty} f(n).$$

4.H. Let $X = R$, $X = B$, and let λ be the Lebesgue measure on B. If $f_n = \chi_{[0,n]}$, then the sequence is monotone increasing to $f = \chi_{[0, +\infty)}$. Although the functions are uniformly bounded by 1 and the integrals of the f_n are all finite, we have

$$\int f \, d\lambda = +\infty.$$

Does the Monotone Convergence Theorem apply?

4.I. Let $X = R$, $X = B$, and λ be Lebesgue measure on X. If $f_n = (1/n) \chi_{[n, +\infty)}$, then the sequence (f_n) is monotone decreasing and converges uniformly to $f = 0$, but

$$0 = \int f \, d\lambda \neq \lim \int f_n \, d\lambda = +\infty.$$

(Hence there is no theorem corresponding to the Monotone Convergence Theorem for a decreasing sequence in M^+.)

4.J.(a) Let $f_n = (1/n) \chi_{[0,n]}$, $f = 0$. Show that the sequence (f_n) converges uniformly to f, but that

$$\int f \, d\lambda \neq \lim \int f_n \, d\lambda.$$

Why does this not contradict the Monotone Convergence Theorem?

Does Fatou's Lemma apply?

(b) Let $g_n = n \chi_{[1/n, 2/n]}$, $g = 0$. Show that

$$\int g \, d\lambda \neq \lim \int g_n \, d\lambda.$$

Does the sequence (g_n) converge uniformly to g? Does the Monotone Convergence Theorem apply? Does Fatou's Lemma apply?

4.K. If (X, X, μ) is a finite measure space, and if (f_n) is a real-valued sequence in $M^+(X, X)$ which converges uniformly to a function f, then f belongs to $M^+(X, X)$, and

$$\int f \, d\mu = \lim \int f_n \, d\mu.$$

4.L. Let X be a finite closed interval $[a, b]$ in R, let X be the collection of Borel sets in X, and let λ be Lebesgue measure on X. If f is a nonnegative continuous function on X, show that

$$\int f \, d\lambda = \int_a^b f(x) \, dx,$$

where the right side denotes the Riemann integral of f. (*Hint:* First establish this equality for a nonnegative **step function**, that is, a linear combination of characteristic functions of intervals.)

4.M. Let $X = [0, +\infty)$, let X be the Borel subsets of X, and let λ be Lebesgue measure on X. If f is a nonnegative continuous function on X, show that

$$\int f \, d\lambda = \lim_{b \to +\infty} \int_0^b f(x) \, dx.$$

Hence, if f is a nonnegative continuous function, the Lebesgue and the improper Riemann integrals coincide.

[The next three exercises deal with the integration of functions which do not belong to M^+. They can be omitted until the next chapter has been read. However, we include them here because they illustrate the restrictions required by Fatou's Lemma.]

4.N. If $f_n = (-1/n)\,\chi_{[0,\,n]}$, then the sequence (f_n) converges uniformly to $f = 0$ on $[0, \infty)$. However, $\int f_n \, d\lambda = -1$ whereas $\int f \, d\lambda = 0$, so

$$\liminf \int f_n \, d\lambda = -1 < 0 = \int f \, d\lambda.$$

Hence Fatou's Lemma 4.8 may not hold unless $f_n \geqslant 0$, even in the presence of uniform convergence.

4.O. Fatou's Lemma has an extension to a case where the f_n take on negative values. Let h be in $M^+(X, X)$, and suppose that $\int h \, d\mu < +\infty$. If (f_n) is a sequence in $M(X, X)$ and if $-h \leqslant f_n$, then

$$\int (\liminf f_n) \, d\mu \leqslant \liminf \int f_n \, d\mu.$$

4.P. Why doesn't Exercise 4.O apply to Exercise 4.N?

4.Q. If $f \in M^+(X, X)$ and

$$\int f \, d\mu < +\infty,$$

then $\mu\{x \in X : f(x) = +\infty\} = 0$. [*Hint:* If $E_n = \{x \in X : f(x) \geqslant n\}$, then $n\chi_{E_n} \leqslant f$.]

4.R. If $f \in M^+(X, X)$ and

$$\int f \, d\mu < +\infty,$$

then the set $N = \{x \in X : f(x) > 0\}$ is σ-finite (that is, there exists a sequence (F_n) in X such that $N \subseteq \bigcup F_n$ and $\mu(F_n) < +\infty$).

4.S. If $f \in M^+(X, X)$ and

$$\int f \, d\mu < +\infty,$$

then for any $\varepsilon > 0$ there exists a set $E \in X$ such that $\mu(E) < +\infty$ and

$$\int f \, d\mu \leqslant \int_E f \, d\mu + \varepsilon.$$

4.T. Suppose that $(f_n) \subset M^+(X, X)$, that (f_n) converges to f, and that

$$\int f \, d\mu = \lim \int f_n \, d\mu < +\infty.$$

Prove that

$$\int_E f \, d\mu = \lim \int_E f_n \, d\mu$$

for each $E \in X$.

4.U. Show that the conclusion of Exercise 4.T may fail if the condition

$$\lim \int f_n \, d\mu < +\infty$$

is dropped.

CHAPTER 5

Integrable Functions

In Definition 4.4 we defined the integral of each function in $M^+ = M^+(X, X)$ with respect to a measure μ and permitted this integral to be $+\infty$. In this chapter we shall discuss the integration of measurable functions which may take on both positive and negative real values. Here it is more convenient to require the values of the functions and the integral to be finite real numbers.

5.1 DEFINITION. The collection $L = L(X, X, \mu)$ of **integrable** (or **summable**) **functions** consists of all real-valued X-measurable functions f defined on X, such that both the positive and negative parts f^+, f^-, of f have finite integrals with respect to μ. In this case, we define the **integral of f with respect to** μ to be

(5.1) $$\int f\, d\mu = \int f^+\, d\mu - \int f^-\, d\mu.$$

If E belongs to X, we define

$$\int_E f\, d\mu = \int_E f^+\, d\mu - \int_E f^-\, d\mu.$$

Although the integral of f is defined to be the difference of the integrals of f^+, f^-, it is easy to see that if $f = f_1 - f_2$ where f_1, f_2 are any nonnegative measurable functions with finite integrals, then

$$\int f\, d\mu = \int f_1\, d\mu - \int f_2\, d\mu.$$

41

In fact, since $f^+ - f^- = f = f_1 - f_2$, it follows that $f^+ + f_2 = f_1 + f^-$. If we apply Corollary 4.7(b), we infer that

$$\int f^+ \, d\mu + \int f_2 \, d\mu = \int f_1 \, d\mu + \int f^- \, d\mu.$$

Since all these terms are finite, we obtain

$$\int f \, d\mu = \int f^+ \, d\mu - \int f^- \, d\mu = \int f_1 \, d\mu - \int f_2 \, d\mu.$$

5.2 LEMMA.   *If f belongs to L and λ is defined on X to **R** by*

(5.2)
$$\lambda(E) = \int_E f \, d\mu,$$

then λ is a charge.

PROOF.   Since f^+ and f^- belong to M^+, Corollary 4.9 implies that the functions λ^+ and λ^-, defined by

$$\lambda^+(E) = \int_E f^+ \, d\mu, \quad \lambda^-(E) = \int_E f^- \, d\mu,$$

are measures on X; they are finite because $f \in L$. Since $\lambda = \lambda^+ - \lambda^-$, it follows that λ is a charge.   Q.E.D.

The function λ defined in (5.2) is frequently called the **indefinite integral of f** (with respect to μ).   Since λ is a charge, if (E_n) is a disjoint sequence in X with union E, then

$$\int_E f \, d\mu = \sum_{n=1}^{\infty} \int_{E_n} f \, d\mu.$$

We refer to this relation by saying that the *indefinite integral of a function in L is countably additive.*

The next result is sometimes referred to as the *property of absolute integrability* of the Lebesgue integral.   The reader will recall that, although the absolute value of a (proper) Riemann integrable function is Riemann integrable, this may no longer be the case for a function which has an improper Riemann integral (for example, consider $f(x) = x^{-1} \sin x$ on the infinite interval $1 \leqslant x < +\infty$).

5.3 THEOREM. *A measurable function f belongs to L if and only if* |f| *belongs to L. In this case*

(5.3) $$\left| \int f \, d\mu \right| \leqslant \int |f| \, d\mu.$$

PROOF. By definition f belongs to L if and only if f^+ and f^- belong to M^+ and have finite integrals. Since $|f|^+ = |f| = f^+ + f^-$ and $|f|^- = 0$, the assertion follows from Lemma 4.5(a) and Corollary 4.7(b). Moreover,

$$\left| \int f \, d\mu \right| = \left| \int f^+ \, d\mu - \int f^- \, d\mu \right|$$
$$\leqslant \int f^+ \, d\mu + \int f^- \, d\mu = \int |f| \, d\mu. \qquad \text{Q.E.D.}$$

5.4 COROLLARY. *If f is measurable, g is integrable, and* $|f| \leqslant |g|$, *then f is integrable, and*

$$\int |f| \, d\mu \leqslant \int |g| \, d\mu.$$

PROOF. This follows from Lemma 4.5(a) and Theorem 5.3. Q.E.D.

We shall now show that *the integral is linear on the space L* in the following sense.

5.5 THEOREM. *A constant multiple* αf *and a sum* $f + g$ *of functions in L belongs to L and*

$$\int \alpha f \, d\mu = \alpha \int f \, d\mu, \quad \int (f + g) \, d\mu = \int f \, d\mu + \int g \, d\mu.$$

PROOF. If $\alpha = 0$, then $\alpha f = 0$ everywhere so that

$$\int \alpha f \, d\mu = 0 = \alpha \int f \, d\mu.$$

If $\alpha > 0$, then $(\alpha f)^+ = \alpha f^+$ and $(\alpha f)^- = \alpha f^-$, whence

$$\int \alpha f \, d\mu = \int \alpha f^+ \, d\mu - \int \alpha f^- \, d\mu$$
$$= \alpha \left\{ \int f^+ \, d\mu - \int f^- \, d\mu \right\} = \alpha \int f \, d\mu.$$

The case $\alpha < 0$ is handled similarly.

If f and g belong to L, then $|f|$ and $|g|$ belong to L. Since $|f + g| \le |f| + |g|$ it follows from Corollaries 4.7 and 5.4 that $f + g$ belongs to L. To establish the desired relation, we observe that

$$f + g = (f^+ + g^+) - (f^- + g^-).$$

Since $f^+ + g^+$ and $f^- + g^-$ are nonnegative integrable functions, it follows from the observation made after Definition 5.1 that

$$\int (f + g)\, d\mu = \int (f^+ + g^+)\, d\mu - \int (f^- + g^-)\, d\mu.$$

If we apply Corollary 4.7(b) and rearrange the terms, we obtain

$$\int (f + g)\, d\mu = \int f^+\, d\mu - \int f^-\, d\mu + \int g^+\, d\mu - \int g^-\, d\mu$$
$$= \int f\, d\mu + \int g\, d\mu. \qquad \text{Q.E.D.}$$

We shall now establish the most important convergence theorem for integrable functions.

5.6 LEBESGUE DOMINATED CONVERGENCE THEOREM. *Let* (f_n) *be a sequence of integrable functions which converges almost everywhere to a real-valued measurable function* f. *If there exists an integrable function* g *such that* $|f_n| \le g$ *for all* n, *then* f *is integrable and*

$$(5.4) \qquad \int f\, d\mu = \lim \int f_n\, d\mu.$$

PROOF. By redefining the functions f_n, f on a set of measure 0 we may assume that the convergence takes place on all of X. It follows from Corollary 5.4 that f is integrable. Since $g + f_n \ge 0$, we can apply Fatou's Lemma 4.8 and Theorem 5.5 to obtain

$$\int g\, d\mu + \int f\, d\mu = \int (g + f)\, d\mu \le \liminf \int (g + f_n)\, d\mu$$
$$= \liminf \left(\int g\, d\mu + \int f_n\, d\mu \right)$$
$$= \int g\, d\mu + \liminf \int f_n\, d\mu.$$

Therefore, it follows that

(5.5) $$\int f \, d\mu \leqslant \lim \inf \int f_n \, d\mu.$$

Since $g - f_n \geqslant 0$, another application of Fatou's Lemma and Theorem 5.5 yields

$$\int g \, d\mu - \int f \, d\mu = \int (g - f) \, d\mu \leqslant \lim \inf \int (g - f_n) \, d\mu$$

$$= \int g \, d\mu - \lim \sup \int f_n \, d\mu,$$

from which it follows that

(5.6) $$\lim \sup \int f_n \leqslant \int f \, d\mu.$$

Combine (5.5) and (5.6) to infer that

$$\int f \, d\mu = \lim \int f_n \, d\mu. \qquad\qquad \text{Q.E.D.}$$

DEPENDENCE ON A PARAMETER

Frequently one needs to consider integrals where the integrand depends on a real parameter. We shall show how the Lebesgue Dominated Convergence Theorem can be used in this connection.

For the remainder of this chapter we shall let f denote a function defined on $X \times [a, b]$ to R and shall assume that the function $x \to f(x, t)$ is X-measurable for each $t \in [a, b]$. Additional hypotheses will be stated explicitly.

5.7 COROLLARY. *Suppose that for some t_0 in $[a, b]$*

(5.7) $$f(x, t_0) = \lim_{t \to t_0} f(x, t)$$

for each $x \in X$, and that there exists an integrable function g on X such that $|f(x, t)| \leqslant g(x)$. Then

$$\int f(x, t_0) \, d\mu(x) = \lim_{t \to t_0} \int f(x, t) \, d\mu(x).$$

PROOF.   Let (t_n) be a sequence in $[a, b]$ which converges to t_0, and apply the Dominated Convergence Theorem to the sequence (f_n) defined by $f_n(x) = f(x, t_n)$ for $x \in X$.   Q.E.D.

5.8 COROLLARY.   *If the function $t \to f(x, t)$ is continuous on $[a, b]$ for each $x \in X$, and if there is an integrable function g on X such that $|f(x, t)| \leqslant g(x)$, then the function F defined by*

$$(5.8) \qquad\qquad F(t) = \int f(x, t)\, d\mu(x)$$

is continuous for t in $[a, b]$.

PROOF.   This is an immediate consequence of Corollary 5.7.   Q.E.D.

5.9 COROLLARY.   *Suppose that for some $t_0 \in [a, b]$, the function $x \to f(x, t_0)$ is integrable on X, that $\partial f/\partial t$ exists on $X \times [a, b]$, and that there exists an integrable function g on X such that*

$$\left| \frac{\partial f}{\partial t}(x, t) \right| \leqslant g(x).$$

Then the function F defined in Corollary 5.8 is differentiable on $[a, b]$ and

$$\frac{dF}{dt}(t) = \frac{d}{dt} \int f(x, t)\, d\mu(x) = \int \frac{\partial f}{\partial t}(x, t)\, d\mu(x).$$

PROOF.   Let t be any point of $[a, b]$.   If (t_n) is a sequence in $[a, b]$ converging to t with $t_n \neq t$, then

$$\frac{\partial f}{\partial t}(x, t) = \lim \frac{f(x, t_n) - f(x, t)}{t_n - t}, \qquad x \in X.$$

Therefore, the function $x \to (\partial f/\partial t)(x, t)$ is measurable.

If $x \in X$ and $t \in [a, b]$, we can apply the Mean Value Theorem (see Reference [1], page 210) to infer the existence of a s_1 between t_0 and t such that

$$f(x, t) - f(x, t_0) = (t - t_0) \frac{\partial f}{\partial t}(x, s_1).$$

Therefore we have

$$|f(x, t)| \leqslant |f(x, t_0)| + |t - t_0|\, g(x),$$

which shows that the function $x \to f(x, t)$ is integrable for each t in $[a, b]$. Hence, if $t_n \neq t$, then

$$\frac{F(t_n) - F(t)}{t_n - t} = \int \frac{f(x, t_n) - f(x, t)}{t_n - t} \, d\mu(x).$$

Since this integrand is dominated by $g(x)$, we may apply the Dominated Convergence Theorem to obtain the stated conclusion.   Q.E.D.

5.10 COROLLARY.   *Under the hypotheses of Corollary 5.8,*

$$\int_a^b F(t) \, dt = \int_a^b \left[\int f(x, t) \, d\mu(x) \right] dt$$

$$= \int \left[\int_a^b f(x, t) \, dt \right] d\mu(x),$$

where the integrals with respect to t are Riemann integrals.

PROOF.   Recall that if φ is continuous on $[a, b]$ then

$$\frac{d}{dt} \int_a^t \varphi(s) \, ds = \varphi(t), \qquad a \leqslant t \leqslant b.$$

Let h be defined on $X \times [a, b]$ by

$$h(x, t) = \int_a^t f(x, s) \, ds.$$

It follows that $(\partial h / \partial t)(x, t) = f(x, t)$. Since this Riemann integral exists, it is the limit of a sequence of Riemann sums; hence the map $x \to h(x, t)$ is measurable for each t.   Moreover, since $|f(x, t)| \leqslant g(x)$, we infer that $|h(x, t)| \leqslant g(x)(b - a)$, so that the function $x \to h(x, t)$ is integrable for each $t \in [a, b]$.   Let H be defined on $[a, b]$ by

$$H(t) = \int h(x, t) \, d\mu(x);$$

it follows from Corollary 5.9 that

$$\frac{dH}{dt}(t) = \int \frac{\partial h}{\partial t}(x, t) \, d\mu(x) = \int f(x, t) \, d\mu(x) = F(t).$$

Therefore we have

$$\int_a^b F(t)\,dt = H(b) - H(a)$$

$$= \int [h(x, b) - h(x, a)]\,d\mu(x)$$

$$= \int \left[\int_a^b f(x, t)\,dt \right] d\mu(x).$$

Q.E.D.

The interchange of the order of (Lebesgue) integrals will be considered in Chapter 10.

EXERCISES

5.A. If $f \in L(X, X, \mu)$ and $a > 0$, show that the set $\{x \in X : |f(x)| \geqslant a\}$ has finite measure. In addition, the set $\{x \in X : f(x) \neq 0\}$ has σ-finite measure (i.e., it is the union of a sequence of measurable sets with finite measure).

5.B. If f is an X-measurable real-valued function and if $f(x) = 0$ for μ-almost all x in X, then $f \in L(X, X, \mu)$ and

$$\int f\,d\mu = 0.$$

5.C. If $f \in L(X, X, \mu)$ and g is an X-measurable real-valued function such that $f(x) = g(x)$ almost everywhere on X, then $g \in L(X, X, \mu)$ and

$$\int f\,d\mu = \int g\,d\mu.$$

5.D. If $f \in L(X, X, \mu)$ and $\varepsilon > 0$, then there exists a measurable simple function φ such that

$$\int |f - \varphi|\,d\mu < \varepsilon.$$

5.E. If $f \in L$ and g is a bounded measurable function, then the product fg also belongs to L.

5.F. If f belongs to L, then it does not follow that f^2 belongs to L.

5.G. Suppose that f is in $L(X, X, \mu)$ and that its indefinite integral is

$$\lambda(E) = \int_E f \, d\mu, \qquad E \in X.$$

Show that $\lambda(E) \geqslant 0$ for all $E \in X$ if and only if $f(x) \geqslant 0$ for almost all $x \in X$. Moreover, $\lambda(E) = 0$ for all E if and only if $f(x) = 0$ for almost all $x \in X$.

5.H. Suppose that f_1 and f_2 are in $L(X, X, \mu)$ and let λ_1 and λ_2 be their indefinite integrals. Show that $\lambda_1(E) = \lambda_2(E)$ for all $E \in X$ if and only if $f_1(x) = f_2(x)$ for almost all x in X.

5.I. If f is a complex-valued function on X such that $\operatorname{Re} f$ and $\operatorname{Im} f$ belong to $L(X, X, \mu)$, we say that f is **integrable** and define

$$\int f \, d\mu = \int \operatorname{Re} f \, d\mu + i \int \operatorname{Im} f \, d\mu.$$

Let f be a complex-valued measurable function. Show that f is integrable if and only if $|f|$ is integrable, in which case

$$\left| \int f \, d\mu \right| \leqslant \int |f| \, d\mu.$$

[*Hint:* If $\int f \, d\mu = r \, e^{i\theta}$ with r, θ real, consider $g(x) = e^{-i\theta} f(x)$.]

5.J. Let (f_n) be a sequence of complex-valued measurable functions which converges to f. If there exists an integrable function g such that $|f_n| \leqslant g$, show that

$$\int f \, d\mu = \lim \int f_n \, d\mu.$$

5.K. Let $X = N$, let X be all subsets of N, and let μ be the counting measure on X. Show that f belongs to $L(X, X, \mu)$ if and only if the series $\sum f(n)$ is absolutely convergent, in which case

$$\int f \, d\mu = \sum_{n=1}^{\infty} f(n).$$

5.L. If (f_n) is a sequence in $L(X, X, \mu)$ which converges uniformly on X to a function f, and if $\mu(X) < +\infty$, then

$$\int f \, d\mu = \lim \int f_n \, d\mu.$$

5.M. Show that the conclusion in the Exercise 5.L may fail if the hypothesis $\mu(X) < +\infty$ is dropped.

5.N. Let $f_n = n \, \chi_{[0, 1/n]}$, where $X = R$, $X = B$, and μ is Lebesgue measure. Show that the condition $|f_n| \leqslant g$ cannot be dropped in the Lebesgue Dominated Convergence Theorem.

5.O. If $f_n \in L(X, X, \mu)$, and if

$$\sum_{n=1}^{\infty} \int |f_n| \, d\mu < +\infty,$$

then the series $\sum f_n(x)$ converges almost everywhere to a function f in $L(X, X, \mu)$. Moreover,

$$\int f \, d\mu = \sum_{n=1}^{\infty} \int f_n \, d\mu.$$

5.P. Let $f_n \in L(X, X, \mu)$, and suppose that (f_n) converges to a function f. Show that if

$$\lim \int |f_n - f| \, d\mu = 0, \quad \text{then} \quad \int |f| \, d\mu = \lim \int |f_n| \, d\mu.$$

5.Q. If $t > 0$, then

$$\int_0^{+\infty} e^{-tx} \, dx = \frac{1}{t}.$$

Moreover, if $t \geqslant a > 0$, then $e^{-tx} \leqslant e^{-ax}$. Use this and Exercise 4.M to justify differentiating under the integral sign and to obtain the formula

$$\int_0^{+\infty} x^n e^{-x} \, dx = n!$$

5.R. Suppose that f is defined on $X \times [a, b]$ to R and that the function $x \to f(t, x)$ is X-measurable for each $t \in [a, b]$. Suppose that for some t_0 and t_1 in $[a, b]$ the function $x \to f(x, t_0)$ is integrable on X, that $(\partial f/\partial t)(x, t_1)$ exists, and that there exists an integrable function g on X such that

$$\left| \frac{f(x, t) - f(x, t_1)}{t - t_1} \right| \leqslant g(x)$$

for $x \in X$, and $t \in [a, b]$, $t \neq t_1$. Then

$$\left[\frac{d}{dt} \int f(x, t) \, d\mu(x) \right]_{t=t_1} = \int \frac{\partial f}{\partial t}(x, t_1) \, d\mu(x).$$

5.S. Suppose the function $x \to f(x, t)$ is X-measurable for each $t \in R$, and the function $t \to f(x, t)$ is continuous on R for each $x \in X$. In addition, suppose that there are integrable functions g, h on X such that $|f(x, t)| \leq g(x)$ and such that the improper Riemann integral

$$\int_{-\infty}^{+\infty} |f(x, t)| \, dt \leq h(x).$$

Show that

$$\int_{-\infty}^{+\infty} \left[\int f(x, t) \, d\mu(x) \right] dt = \int \left[\int_{-\infty}^{+\infty} f(x, t) \, dt \right] d\mu(x),$$

where the integrals with respect to t are improper Riemann integrals.

5.T. Let f be an X-measurable function on X to R. For $n \in N$, let (f_n) be the sequence of truncates of f (see Exercise 2.K). If f is integrable with respect to μ, then

$$\int f \, d\mu = \lim \int f_n \, d\mu.$$

Conversely, if

$$\sup \int |f_n| \, d\mu < +\infty,$$

then f is integrable.

CHAPTER 6

The Lebesgue spaces L_p

It is often useful to impose the structure of a Banach space on the set of all integrable functions on a measure space (X, X, μ). In addition, we shall introduce the L_p, $1 \leqslant p \leqslant \infty$, spaces which occur frequently in analysis. Aside from the intrinsic importance of these spaces, we examine them here partly to indicate applications of some of the results in the earlier sections.

6.1 DEFINITION. If V is a real linear ($=$ vector) space, then a real-valued function N on V is said to be a **norm** for V in case it satisfies

(i) $N(v) \geqslant 0$ for all $v \in V$;

(ii) $N(v) = 0$ if and only if $v = 0$;

(iii) $N(\alpha v) = |\alpha| N(v)$ for all $v \in V$ and real α;

(iv) $N(u + v) \leqslant N(u) + N(v)$ for all $u, v \in V$.

If condition (ii) is dropped, the function N is said to be a **semi-norm** or a **pseudo-norm** for V. A **normed linear space** is a linear space V together with a norm for V.

6.2 EXAMPLES. (a) The absolute value function yields a norm for the real numbers.

(b) The linear space R^n of n-tuples of real numbers can be normed by defining

$$N_1(u_1, \ldots, u_n) = |u_1| + \cdots + |u_n|,$$
$$N_p(u_1, \ldots, u_n) = \{|u_1|^p + \cdots + |u_n|^p\}^{1/p}, \quad p \geqslant 1,$$
$$N_\infty(u_1, \ldots, u_n) = \sup \{|u_1|, \ldots, |u_n|\}.$$

It is easy to check that N_1 and N_∞ are norms and that N_p satisfies (i), (ii), (iii). It is a consequence of Minkowski's Inequality, which will be proved subsequently, that N_p satisfies (iv).

(c) The linear space l_1 of all real-valued sequences $u = (u_n)$ such that $N_1(u) = \sum |u_n| < +\infty$ is a normed linear space under N_1. Similarly, if $1 \leqslant p < \infty$, the collection l_p of all sequences such that $N_p(u) = \{\sum |u_n|^p\}^{1/p} < +\infty$ is normed by N_p.

(d) The collection $B(X)$ of all bounded real-valued functions on X is normed by

$$N(f) = \sup\{|f(x)| : x \in X\}.$$

In particular, the linear space of continuous functions on $X = [a, b]$ is normed.

All the preceding examples have been proper norms on a linear space. There are also semi-norms on a linear space that are of interest. The following are some examples.

6.3 EXAMPLES.   (a) On the space R^n, consider the semi-norm

$$N_0(u_1, \ldots, u_n) = \sup\{|u_2|, \ldots, |u_n|\}.$$

Here $N_0(u_1, \ldots, u_n) = 0$ if and only if $u_2 = \cdots = u_n = 0$.

(b) On the linear space $C[0, 1]$ of continuous functions on $[0, 1]$ to R, define the semi-norm

$$N_0(f) = \sup\{|f(x)| : 0 \leqslant x \leqslant \tfrac{1}{2}\}.$$

Here $N_0(f) = 0$ if and only if $f(x)$ vanishes for $0 \leqslant x \leqslant \tfrac{1}{2}$.

(c) On the linear space of functions on $[a, b]$ to R which have continuous derivatives, consider the semi-norm

$$N_0(f) = \sup\{|f'(x)| : a \leqslant x \leqslant b\}.$$

Here $N_0(f) = 0$ if and only if f is constant on $[a, b]$.

6.4 DEFINITION.   Let $(X, \boldsymbol{X}, \mu)$ be a measure space.   If f belongs to $L(X, \boldsymbol{X}, \mu)$, we define

$$N_\mu(f) = \int |f|\, d\mu.$$

It will be shown that N_μ is a semi-norm on the space $L(X, \boldsymbol{X}, \mu)$.

6.5 LEMMA.   *The space $L(X, \mathbf{X}, \mu)$ is a linear space under the operations defined by*

$$(f + g)(x) = f(x) + g(x), \qquad (\alpha f)(x) = \alpha f(x), \quad x \in X,$$

and N_μ is a semi-norm on $L(X, \mathbf{X}, \mu)$.   Moreover, $N_\mu(f) = 0$ if and only if $f(x) = 0$ for μ-almost all x in X.

PROOF.   It was seen in Theorem 5.5 that $L = L(X, \mathbf{X}, \mu)$ is a linear space under the indicated operations.   It is clear that $N_\mu(f) \geqslant 0$ for $f \in L$, and that

$$N_\mu(\alpha f) = \int |\alpha f| \, d\mu = |\alpha| \int |f| \, d\mu = |\alpha| N_\mu(f).$$

Moreover, it follows from the Triangle Inequality that

$$N_\mu(f + g) = \int |f + g| \, d\mu \leqslant \int (|f| + |g|) \, d\mu$$
$$= \int |f| \, d\mu + \int |g| \, d\mu = N_\mu(f) + N_\mu(g).$$

Hence N_μ is a semi-norm on L, and it follows from Corollary 4.10 that $N_\mu(f) = 0$ if and only if $f(x) = 0$ for almost all x.   Q.E.D.

In order to make $L(X, \mathbf{X}, \mu)$ into a normed linear space, we shall identify two functions that are equal almost everywhere; that is, we use equivalence classes of functions instead of functions.

6.6 DEFINITION.   Two functions in $L = L(X, \mathbf{X}, \mu)$ are said to be **μ-equivalent** if they are equal μ-almost everywhere.   The **equivalence class determined by** f in L is sometimes denoted by $[f]$ and consists of the set of all functions in L which are μ-equivalent to f.   The **Lebesgue space** $L_1 = L_1(X, \mathbf{X}, \mu)$ consists of all μ-equivalence classes in L.   If $[f]$ belongs to L_1, we define its **norm** by

(6.1) $$\|[f]\|_1 = \int |f| \, d\mu.$$

6.7 THEOREM.   *The Lebesgue space $L_1(X, \mathbf{X}, \mu)$ is a normed linear space.*

PROOF. It is understood, of course, that the vector operations in L_1 are defined by

$$\alpha[f] = [\alpha f], \quad [f] + [g] = [f + g],$$

and that the zero element of L_1 is $[0]$. We shall check only that equation (6.1) gives a norm on L_1. Certainly $\|[f]\|_1 \geqslant 0$ and $\|[0]\|_1 = 0$. Moreover, if $\|[f]\|_1 = 0$ then

$$\int |f| \, d\mu = 0,$$

so $f(x) = 0$ for μ-almost all x. Hence $[f] = [0]$. Finally, it is easily seen that properties (iii) and (iv) of Definition 6.1 are satisfied. Therefore $\| \ \|_1$ yields a norm on L_1.   Q.E.D.

It should always be remembered that the elements of L_1 are actually equivalence classes of functions in L. However, it is both convenient and customary to regard these elements as being functions, and we shall subsequently do so. Thus we shall make reference to the equivalence class $[f]$ by referring to "the element f of L_1," and we shall write $\|f\|_1$ in place of $\|[f]\|_1$.

THE SPACES L_p, $1 \leqslant p < +\infty$

We now wish to consider a family of related normed linear spaces of equivalence classes of measurable functions.

6.8 DEFINITION. If $1 \leqslant p < \infty$, the space $L_p = L_p(X, X, \mu)$ consists of all μ-equivalence classes of X-measurable real-valued functions f for which $|f|^p$ has finite integral with respect to μ over X. Two functions are μ-**equivalent** if they are equal μ-almost everywhere. We set

$$(6.3) \qquad \|f\|_p = \left\{ \int |f|^p \, d\mu \right\}^{1/p}.$$

If $p = 1$, this reduces to the norm introduced previously on the space L_1 of equivalence classes of integrable functions. We shall show subsequently that if $1 \leqslant p < \infty$, then L_p is a normed linear space under (6.3), and is complete under this norm; thus L_p is a Banach space. It is understood that the vector operations between the equivalence classes in L_p are defined pointwise: the sum of the equivalence classes containing f and g is the equivalence class containing $f + g$ and similarly for the product cf.

In the special case where μ is the counting measure on all subsets of N, the L_p-spaces can be identified with the sequence spaces l_p of Example 6.2(c). In this case, each equivalence class contains one element. It is frequently enlightening to interpret assertions about general L_p-spaces by considering the somewhat simpler l_p-spaces.

In order to establish that (6.3) yields a norm on L_p, we shall need the following basic inequality.

6.9 HÖLDER'S INEQUALITY. *Let* $f \in L_p$ *and* $g \in L_q$ *where* $p > 1$ *and* $(1/p) + (1/q) = 1$. *Then* $fg \in L_1$ *and* $\|fg\|_1 \leqslant \|f\|_p \|g\|_q$.

PROOF. Let α be a real number satisfying $0 < \alpha < 1$, and consider the function φ defined for $t \geqslant 0$ by

$$\varphi(t) = \alpha t - t^\alpha.$$

It is easy to check that $\varphi'(t) < 0$ for $0 < t < 1$ and $\varphi'(t) > 0$ for $t > 1$. It follows from the Mean Value Theorem of calculus that $\varphi(t) \geqslant \varphi(1)$ and that $\varphi(t) = \varphi(1)$, if and only if $t = 1$. Therefore we have

$$t^\alpha \leqslant \alpha t + (1 - \alpha), \qquad t \geqslant 0.$$

If a, b are nonnegative, and if we let $t = a/b$ and multiply by b, we obtain the inequality

$$a^\alpha b^{1-\alpha} \leqslant \alpha a + (1 - \alpha)b,$$

where equality holds if and only if $a = b$.

Now let p and q satisfy $1 < p < \infty$ and $(1/p) + (1/q) = 1$ and take $\alpha = 1/p$. It follows that if A, B are any nonnegative real numbers, then

(6.4) $$AB \leqslant \frac{A^p}{p} + \frac{B^q}{q}.$$

and that the equality holds if and only if $A^p = B^q$.

Suppose that $f \in L_p$ and $g \in L_q$, and that $\|f\|_p \neq 0$ and $\|g\|_q \neq 0$.

The product fg is measurable and (6.4) with $A = |f(x)|/\|f\|_p$ and $B = |g(x)|/\|g\|_q$ implies that

$$\frac{|f(x)g(x)|}{\|f\|_p\,\|g\|_q} \leqslant \frac{|f(x)|^p}{p\|f\|_p{}^p} + \frac{|g(x)|^q}{q\|g\|_q{}^q}\,.$$

Since both of the terms on the right are integrable, it follows from Corollary 5.4 and Theorem 5.5 that fg is integrable. Moreover, on integrating we obtain

$$\frac{\|fg\|_1}{\|f\|_p\|g\|_q} \leqslant \frac{1}{p} + \frac{1}{q} = 1.$$

which is Hölder's Inequality.   Q.E.D.

Hölder's Inequality implies that the product of a function in L_p and a function in L_q is integrable when $p > 1$ and q satisfies the relation $(1/p) + (1/q) = 1$ or, equivalently, when $p + q = pq$. Two numbers satisfying this relation are said to be **conjugate indices**. It will be noted that $p = 2$ is the only **self-conjugate** index. Thus the product of two functions in L_2 is integrable.

6.10 CAUCHY–BUNYAKOVSKIĬ–SCHWARZ INEQUALITY. *If f and g belong to L_2, then fg is integrable and*

$$(6.5) \qquad \left| \int fg \, d\mu \right| \leqslant \int |fg| \, d\mu \leqslant \|f\|_2 \, \|g\|_2.$$

6.11 MINKOWSKI'S INEQUALITY. *If f and h belong to L_p, $p \geqslant 1$, then $f + h$ belongs to L_p and*

$$(6.6) \qquad \|f + h\|_p \leqslant \|f\|_p + \|h\|_p.$$

PROOF. The case $p = 1$ has already been treated, so we suppose $p > 1$. The sum $f + h$ is evidently measurable. Since

$$|f + h|^p \leqslant [2 \sup \{|f|,|h|\}]^p \leqslant 2^p\{|f|^p + |h|^p\}$$

it follows from Corollary 5.4 and Theorem 5.5 that $f + h \in L_p$. Moreover,

$$(6.7) \quad |f + h|^p = |f + h|\,|f + h|^{p-1} \leqslant |f|\,|f + h|^{p-1} + |h|\,|f + h|^{p-1}.$$

Since $f + h \in L_p$, then $|f + h|^p \in L_1$; since $p = (p - 1)q$ it follows that

$|f + h|^{p-1} \in L_q$.   Hence we can apply Hölder's Inequality to infer that

$$\int |f| \, |f + h|^{p-1} \, d\mu \leqslant \|f\|_p \left\{ \int |f + h|^{(p-1)q} \, d\mu \right\}^{1/q}$$
$$= \|f\|_p \, \|f + h\|_p^{p/q}.$$

If we treat the second term on the right in (6.7) similarly, we obtain

$$\|f + h\|_p^p \leqslant \|f\|_p \, \|f + h\|_p^{p/q} + \|h\|_p \, \|f + h\|_p^{p/q}$$
$$= \{\|f\|_p + \|h\|_p\} \, \|f + h\|_p^{p/q}.$$

If $A = \|f + h\|_p = 0$, then equation (6.6) is trivial. If $A \neq 0$, we can divide the above inequality by $A^{p/q}$; since $p - p/q = 1$, we obtain Minkowski's Inequality.   Q.E.D.

It is readily seen that the space L_p is a linear space and that formula (6.3) defines a norm on L_p.   The only nontrivial thing to be checked here is the inequality 6.1(iv) and this is Minkowski's Inequality.   We shall now show that L_p is complete under this norm in the following sense.

6.12 DEFINITION.   A sequence (f_n) in L_p is a **Cauchy sequence** in L_p if for every positive number ε there exists an $M(\varepsilon)$ such that if $m, n \geqslant M(\varepsilon)$, then $\|f_m - f_n\|_p < \varepsilon$.   A sequence (f_n) in L_p is **convergent to** f in L_p if for every positive number ε there exists an $N(\varepsilon)$ such that if $n \geqslant N(\varepsilon)$, then $\|f - f_n\|_p < \varepsilon$.   A normed linear space is **complete** if every Cauchy sequence converges to some element of the space.

6.13 LEMMA.   *If the sequence (f_n) converges to f in L_p, then it is a Cauchy sequence.*

PROOF.   If $m, n \geqslant N(\varepsilon/2)$, then

$$\|f - f_m\|_p < \frac{\varepsilon}{2}, \qquad \|f - f_n\|_p < \frac{\varepsilon}{2}.$$

Hence we have

$$\|f_m - f_n\|_p \leqslant \|f_m - f\|_p + \|f - f_n\|_p < \varepsilon.$$ Q.E.D.

We shall now show that every Cauchy sequence in L_p converges in L_p to an element.   This result is sometimes called the Riesz–Fischer Theorem.

6.14 COMPLETENESS THEOREM. *If $1 \leqslant p < \infty$, then the space L_p is a complete normed linear space under the norm*

$$\|f\|_p = \left\{ \int |f|^p \, d\mu \right\}^{1/p}.$$

PROOF. It has been stated that L_p is a normed linear space. To establish the completeness of L_p, let (f_n) be a Cauchy sequence relative to the norm $\| \ \|_p$. Hence, if $\varepsilon > 0$ there exists an $M(\varepsilon)$ such that if $m, n \geqslant M(\varepsilon)$, then

$$(6.8) \qquad \int |f_m - f_n|^p \, d\mu = \|f_m - f_n\|_p^p < \varepsilon^p.$$

There exists a subsequence (g_k) of (f_n) such that $\|g_{k+1} - g_k\|_p < 2^{-k}$ for $k \in N$. Define g by

$$(6.9) \qquad g(x) = |g_1(x)| + \sum_{k=1}^{\infty} |g_{k+1}(x) - g_k(x)|,$$

so that g is in $M^+(X, X)$. By Fatou's Lemma, we have

$$\int |g|^p \, d\mu \leqslant \liminf_{n \to \infty} \int \left\{ |g_1| + \sum_{k=1}^{n} |g_{k+1} - g_k| \right\}^p d\mu.$$

Take the pth root of both sides and apply Minkowski's Inequality to obtain

$$\left\{ \int |g|^p \, d\mu \right\}^{1/p} \leqslant \liminf_{n \to \infty} \left\{ \|g_1\|_p + \sum_{k=1}^{n} \|g_{k+1} - g_k\|_p \right\}$$
$$\leqslant \|g_1\|_p + 1.$$

Hence, if $E = \{x \in X : g(x) < +\infty\}$, then $E \in X$ and $\mu(X \setminus E) = 0$. Therefore, the series in (6.9) converges almost everywhere and $g \chi_E$ belongs to L_p.

We now define f on X by

$$f(x) = g_1(x) + \sum_{k=1}^{\infty} \{g_{k+1}(x) - g_k(x)\}, \qquad x \in E,$$
$$= 0, \qquad x \notin E.$$

Since $|g_k| \leqslant |g_1| + \sum_{j=1}^{k-1} |g_{j+1} - g_j| \leqslant g$ and since (g_k) converges almost everywhere to f, the Dominated Convergence Theorem 5.6

implies that $f \in L_p$. Since $|f - g_k|^p \leqslant 2^p g^p$, we infer from the Dominated Convergence Theorem that $0 = \lim \|f - g_k\|_p$, so that (g_k) converges in L_p to f.

In view of (6.8), if $m \geqslant M(\varepsilon)$ and k is sufficiently large, then

$$\int |f_m - g_k|^p \, d\mu < \varepsilon^p.$$

Apply Fatou's Lemma to conclude that

$$\int |f_m - f|^p \, d\mu \leqslant \liminf_{k \to \infty} \int |f_m - g_k|^p \, d\mu \leqslant \varepsilon^p,$$

whenever $m \geqslant M(\varepsilon)$. This proves that the sequence (f_n) converges to f in the norm of L_p.   Q.E.D.

A complete normed linear space is usually called a **Banach space** Thus the preceding theorem could be formulated: *the space L_p is a Banach space under the norm given in* (6.3).

THE SPACE L_∞

We shall now introduce a space which is related to the L_p-spaces.

6.15 DEFINITION. The space $L_\infty = L_\infty(X, X, \mu)$ consists of all the equivalence classes of X-measurable real-valued functions which are almost everywhere bounded, two functions being equivalent when they are equal μ-almost everywhere. If $f \in L_\infty$ and $N \in X$ with $\mu(N) = 0$, we define

$$S(N) = \sup \{|f(x)| : x \notin N\}$$

and

(6.10) $\|f\|_\infty = \inf \{S(N) : N \in X, \mu(N) = 0\}.$

An element of L_∞ is called an **essentially bounded function.**

It follows (see Exercise 6.T) that if $f \in L_\infty$, then $|f(x)| \leqslant \|f\|_\infty$ for almost all x. Moreover, if $A < \|f\|_\infty$, then there exists a set E with positive measure such that $|f(x)| \geqslant A$ for $x \in E$. It is also clear that the norm in (6.10) is well-defined on L_∞.

6.16 THEOREM. *The space L_∞ is a complete normed linear space under the norm given by formula (6.10).*

PROOF. It is clear that L_∞ is a linear space and that $\|f\|_\infty \geq 0$, $\|0\|_\infty = 0$, and $\|\alpha f\|_\infty = |\alpha| \|f\|_\infty$. If $\|f\|_\infty = 0$, then there exists a set $N_k \in X$ with $\mu(N_k) = 0$ such that $|f(x)| \leq 1/k$ for $x \notin N_k$. If we put $N = \bigcup_{k=1}^\infty N_k$, then $N \in X$, $\mu(N) = 0$, and $|f(x)| = 0$ for $x \notin N$. Therefore, $f(x) = 0$ for almost all x.

If $f, g \in L_\infty$, there exist sets N_1, N_2 in X with $\mu(N_1) = \mu(N_2) = 0$ such that

$$|f(x)| \leq \|f\|_\infty \quad \text{for} \quad x \notin N_1,$$
$$|g(x)| \leq \|g\|_\infty \quad \text{for} \quad x \notin N_2.$$

Therefore $|f(x) + g(x)| \leq \|f\|_\infty + \|g\|_\infty$ for $x \notin (N_1 \cup N_2)$, from which it follows that $\|f + g\|_\infty \leq \|f\|_\infty + \|g\|_\infty$.

It remains to prove that L_∞ is complete. Let (f_n) be a Cauchy sequence in L_∞, and let M be a set in X with $\mu(M) = 0$, such that $|f_n(x)| \leq \|f_n\|_\infty$ for $x \notin M$, $n = 1, 2, \ldots$, and also such that $|f_n(x) - f_m(x)| \leq \|f_n - f_m\|_\infty$ for all $x \notin M$, n, $m = 1, 2, \ldots$. Then the sequence (f_n) is uniformly convergent on $X \setminus M$, and we let

$$f(x) = \lim f_n(x), \quad x \notin M,$$
$$= 0, \quad x \in M.$$

It follows that f is measurable, and it is easily seen that $\|f_n - f\|_\infty \to 0$. Hence L_∞ is complete.   Q.E.D.

EXERCISES

6.A. Let $C[0, 1]$ be the linear space of continuous functions on $[0, 1]$ to R. Define N_0 for f in $C[0, 1]$ by $N_0(f) = |f(0)|$. Show that N_0 is a semi-norm on $C[0, 1]$.

6.B. Let $C[0, 1]$ be as before and define N_1 for f in $C[0, 1]$ to be the Riemann integral of $|f|$ over $[0, 1]$. Show that N_1 defines a norm on $C[0, 1]$. If f_n is defined for $n \geq 1$ to be equal to 0 for $0 \leq x \leq (1 - 1/n)/2$, to be equal to 1 for $\frac{1}{2} \leq x \leq 1$, and to be linear for $(1 - 1/n)/2 \leq x \leq \frac{1}{2}$, show that (f_n) is a Cauchy sequence, but that it does not converge relative to N_1 to an element of $C[0, 1]$.

6.C. Let N be a norm on a linear space V and let d be defined for $u, v \in V$ by $d(u, v) = N(u - v)$. Show that d is a metric on V; that is, (i) $d(u, v) \geqslant 0$ for all $u, v \in V$; (ii) $d(u, v) = 0$ if and only if $u = v$; (iii) $d(u, v) = d(v, u)$; (iv) $d(u, v) \leqslant d(u, w) + d(w, v)$.

6.D. If $f \in L_1(X, X, \mu)$ and $\varepsilon > 0$, then there exists a simple X-measurable function φ such that $\|f - \varphi\|_1 < \varepsilon$. Extend this to L_p, $1 \leqslant p < \infty$. Is this true for L_∞?

6.E. If $f \in L_p$, $1 \leqslant p < \infty$, and if $E = \{x \in X : |f(x)| \neq 0\}$, then E is σ-finite.

6.F. If $f \in L_p$ and if $E_n = \{x \in X : |f(x)| \geqslant n\}$, then $\mu(E_n) \to 0$ as $n \to \infty$.

6.G. Let $X = N$, and let μ be the counting measure on N. If f is defined on N by $f(n) = 1/n$, then f does not belong to L_1, but it does belong to L_p for $1 < p \leqslant \infty$. [Alternatively, let $X = R$, $X = B$, and let μ be Lebesgue measure and define $g(x) = 0$ for $x < 1$ and $g(x) = 1/x$ for $x \geqslant 1$.]

6.H. Let $X = N$, and let λ be the measure on N which has measure $1/n^2$ at the point n. (More precisely $\lambda(E) = \sum \{1/n^2 : n \in E\}$.) Show that $\lambda(X) < +\infty$. Let f be defined on X by $f(n) = \sqrt{n}$. Show that $f \in L_p$ if and only if $1 \leqslant p < 2$. [For a similar example, let $X = (0, 1)$ with Lebesgue measure, and consider $g(x) = 1/\sqrt{x}$.]

6.I. Modify the Exercise 6.H to obtain a function on a finite measure space which belongs to L_p if and only if $1 \leqslant p < p_0$.

6.J. Let (X, X, μ) be a finite measure space. If f is X-measurable, let $E_n = \{x \in X : (n - 1) \leqslant |f(x)| < n\}$. Show that $f \in L_1$ if and only if

$$\sum_{n=1}^{\infty} n \, \mu(E_n) < +\infty.$$

More generally, $f \in L_p$ for $1 \leqslant p < \infty$, if and only if

$$\sum_{n=1}^{\infty} n^p \, \mu(E_n) < +\infty.$$

6.K. If (X, X, μ) is a finite measure space and $f \in L_p$, then $f \in L_r$ for $1 \leqslant r \leqslant p$. (*Hint:* Use Exercise 6.J or the inequality $|f|^r \leqslant 1 +$

$|f|^p$.) Apply Hölder's Inequality to $|f|^r$ in $L_{p/r}$ and $g = 1$ to obtain the inequality

$$\|f\|_r \leqslant \|f\|_p \, \mu(X)^s,$$

where $s = (1/r) - (1/p)$. Therefore, if $\mu(X) = 1$, then $\|f\|_r \leqslant \|f\|_p$.

6.L. Suppose that $X = N$ and μ is the counting measure on N. If $f \in L_p$, then $f \in L_s$ with $1 \leqslant p \leqslant s < \infty$, and $\|f\|_s \leqslant \|f\|_p$.

6.M. Let $X = (0, \infty)$, let μ be Lebesgue measure on X, and let $f(x) = x^{-1/2}(1 + |\log x|)^{-1}$. Then $f \in L_p$ if and only if $p = 2$.

6.N. Let (X, X, μ) be any measure space and let f belong to both L_{p_1} and L_{p_2}, with $1 \leqslant p_1 < p_2 < \infty$. Prove that $f \in L_p$ for any value of p such that $p_1 \leqslant p \leqslant p_2$.

6.O. Let $1 < p < \infty$, and let $(1/p) + (1/q) = 1$. It follows from Hölder's Inequality that if $f \in L_p$, then

$$\left| \int fg \, d\mu \right| \leqslant \|f\|_p$$

for all $g \in L_q$ such that $\|g\|_q \leqslant 1$. If $f \neq 0$, define g_0 on X by $g_0(x) = c[\operatorname{signum} f(x)]|f(x)|^{p-1}$, where $c = (\|f\|_p)^{-p/q}$. Show that $g_0 \in L_q$, that $\|g_0\|_q = 1$, and that

$$\left| \int fg_0 \, d\mu \right| = \|f\|_p.$$

6.P. Let $f \in L_p(X, X, \mu)$, $1 \leqslant p < \infty$, and let $\varepsilon > 0$. Show that there exists a set $E_\varepsilon \in X$ with $\mu(E_\varepsilon) < +\infty$ such that if $F \in X$ and $F \cap E_\varepsilon = \emptyset$, then $\|f\chi_F\|_p < \varepsilon$.

6.Q. Let $f_n \in L_p(X, X, \mu)$, $1 \leqslant p < \infty$, and let β_n be defined for $E \in X$ by

$$\beta_n(E) = \left\{ \int_E |f_n|^p \, d\mu \right\}^{1/p}.$$

Show that $|\beta_n(E) - \beta_m(E)| \leqslant \|f_n - f_m\|_p$. Hence, if (f_n) is a Cauchy sequence in L_p, then $\lim \beta_n(E)$ exists for each $E \in X$.

6.R. Let f_n, β_n be as in Exercise 6.Q. If (f_n) is a Cauchy sequence and $\varepsilon > 0$, then there exists a set $E_\varepsilon \in X$ with $\mu(E_\varepsilon) < +\infty$ such that if $F \in X$ and $F \cap E_\varepsilon = \emptyset$, then $\beta_n(F) < \varepsilon$ for all $n \in N$.

6.S. Let f_n, β_n be as in the Exercise 6.R, and suppose that (f_n) is a Cauchy sequence. If $\varepsilon > 0$, then there exists a $\delta(\varepsilon) > 0$ such that if $E \in X$ and $\mu(E) < \delta(\varepsilon)$, then $\beta_n(E) < \varepsilon$ for all $n \in N$. (*Hint:* Use Corollary 4.11.)

6.T. If $f \in L_\infty(X, X, \mu)$, then $|f(x)| \leqslant \|f\|_\infty$ for almost all x. Moreover, if $A < \|f\|_\infty$, then there exists a set $E \in X$ with $\mu(E) > 0$ such that $|f(x)| > A$ for all $x \in E$.

6.U. If $f \in L_p$, $1 \leqslant p \leqslant \infty$, and $g \in L_\infty$, then the product $fg \in L_p$ and $\|fg\|_p \leqslant \|f\|_p \|g\|_\infty$.

6.V. The space $L_\infty(X, X, \mu)$ is contained in $L_1(X, X, \mu)$ if and only if $\mu(X) < \infty$. If $\mu(X) = 1$ and $f \in L_\infty$, then

$$\|f\|_\infty = \lim_{p \to \infty} \|f\|_p.$$

CHAPTER 7

Modes of Convergence

We have already had occasion to mention four types of convergence of a sequence of measurable functions: pointwise convergence, almost everywhere convergence, uniform convergence, and convergence in L_p. There are two other notions of convergence that are of importance in dealing with measurable functions. We shall introduce these in this chapter and give interrelations between the various modes.

For convenience, we shall restate the definitions. **In this chapter we shall consider only real-valued functions defined on a fixed measure space** (X, X, μ). In some applications it is necessary to consider extended real-valued functions, but this can usually be done by modifying the present discussion. In addition we shall limit our attention to L_p for $1 \leqslant p < \infty$, since the convergence L_∞ requires a special examination which is usually quite direct. Thus it will be understood that p is limited to these values.

The sequence (f_n) **converges uniformly** to f if for every $\varepsilon > 0$ there exists a natural number $N(\varepsilon)$ such that if $n \geqslant N(\varepsilon)$ and $x \in X$, then $|f_n(x) - f(x)| < \varepsilon$.

The sequence (f_n) **converges pointwise** to f if for every $\varepsilon > 0$ and $x \in X$ there is a natural number $N(\varepsilon, x)$, such that if $n \geqslant N(\varepsilon, x)$, then $|f_n(x) - f(x)| < \varepsilon$.

The sequence (f_n) **converges almost everywhere** to f if there exists a set M in X with $\mu(M) = 0$ such that for every $\varepsilon > 0$ and $x \in X \setminus M$

there exists a natural number $N(\varepsilon, x)$, such that if $n \geq N(\varepsilon, x)$, then $|f_n(x) - f(x)| < \varepsilon$.

It is obvious that uniform convergence implies pointwise convergence, that pointwise convergence implies almost everywhere convergence, and it is easily seen that the reverse implications do not hold. (Of course, if X consists of only a finite number of points, then pointwise convergence implies uniform convergence; if the only set with measure zero is the empty set, then almost everywhere convergence implies pointwise convergence.)

CONVERGENCE IN L_p

We now recall the notion of convergence in L_p, which was introduced in Chapter 6. We remark that an element in L_p is an equivalence class of functions which are real-valued and whose pth powers are integrable. However, by exercising some caution, we may regard an element of L_p as being a real-valued measurable function.

A sequence (f_n) in $L_p = L_p(X, X, \mu)$ **converges in** L_p to $f \in L_p$, if for every $\varepsilon > 0$ there exists a natural number $N(\varepsilon)$ such that if $n \geq N(\varepsilon)$, then

$$\|f_n - f\|_p = \left\{ \int |f_n - f|^p \, d\mu \right\}^{1/p} < \varepsilon.$$

In this case, we sometimes say that the sequence (f_n) **converges to f in mean (of order p)**.

A sequence (f_n) in L_p is said to be **Cauchy in** L_p, if for every $\varepsilon > 0$ there exists a natural number $N(\varepsilon)$ such that if $m, n \geq N(\varepsilon)$, then

$$\|f_m - f_n\|_p = \left\{ \int |f_m - f_n|^p \, d\mu \right\}^{1/p} < \varepsilon.$$

We have seen in Theorem 6.14 that if (f_n) is Cauchy in L_p, then there exists an $f \in L_p$ such that (f_n) converges in L_p to f.

The relationship between convergence in L_p and the other modes of convergence that we have introduced is not so close. It is possible (see Exercise 7.A) for a sequence (f_n) in L_p to converge uniformly on X (and therefore pointwise and almost everywhere) to a function f in L_p, but not converge in L_p. However, if $\mu(X) < +\infty$, this cannot be the case.

7.1 THEOREM. *Suppose that $\mu(X) < +\infty$ and that (f_n) is a sequence in L_p which converges uniformly on X to f. Then f belongs to L_p and the sequence (f_n) converges in L_p to f.*

PROOF. Let $\varepsilon > 0$ and let $N(\varepsilon)$ be such that $|f_n(x) - f(x)| < \varepsilon$ whenever $n \geqslant N(\varepsilon)$ and $x \in X$. If $n \geqslant N(\varepsilon)$, then

$$
(7.1) \qquad \|f_n - f\|_p = \left\{ \int |f_n(x) - f(x)|^p \, d\mu \right\}^{1/p}
$$
$$
\leqslant \left\{ \int \varepsilon^p \, d\mu \right\}^{1/p} = \varepsilon \mu(X)^{1/p},
$$

so that (f_n) converges in L_p to f. Q.E.D.

It is possible (see Exercise 7.B) for a sequence (f_n) in L_p to converge pointwise (and therefore almost everywhere) to a function f in L_p, but not converge in L_p even when $\mu(X) < +\infty$. However, if the sequence is dominated by a function in L_p, then the L_p convergence does take place.

7.2 THEOREM. *Let (f_n) be a sequence in L_p which converges almost everywhere to a measurable function f. If there exists a g in L_p such that*

$$
(7.2) \qquad |f_n(x)| \leqslant g(x), \qquad x \in X, \qquad n \in N,
$$

then f belongs to L_p and (f_n) converges in L_p to f.

PROOF. In view of inequality (7.2), it follows that $|f(x)| \leqslant g(x)$ almost everywhere. Since $g \in L_p$, it follows from Corollary 5.4 that $f \in L_p$. Now

$$
|f_n(x) - f(x)|^p \leqslant [2 \, g(x)]^p, \text{ a.e.,}
$$

and since $\lim |f_n(x) - f(x)|^p = 0$, a.e., and $2^p g^p$ belongs to L_1, it follows from the Lebesgue Dominated Convergence Theorem 5.6 that

$$
\lim \int |f_n - f|^p \, d\mu = 0.
$$

Therefore (f_n) converges in L_p to f. Q.E.D.

7.3   COROLLARY.   *Let $\mu(X) < +\infty$, and let (f_n) be a sequence in L_p which converges almost everywhere to a measurable function f.   If there exists a constant K such that*

(7.3) $|f_n(x)| \leqslant K, \quad x \in X, \quad n \in N,$

then f belongs to L_p and (f_n) converges in L_p to f.

PROOF.   If $\mu(X) < +\infty$, the constant functions belong to L_p.   Q.E.D.

It might be suspected that convergence in L_p implies almost everywhere convergence, but this is not the case.   In fact, we shall give an example of sequence (f_n) which converges in L_p to a function f, but such that $(f_n(x))$ does not converge to $f(x)$ for any x in $X(!)$

7.4   EXAMPLE.   Let $X = [0, 1]$, $X = B$, and let λ be Lebesgue measure.   We shall consider the intervals $[0, 1]$, $[0, \frac{1}{2}]$, $[\frac{1}{2}, 1]$, $[0, \frac{1}{3}]$, $[\frac{1}{3}, \frac{2}{3}]$, $[\frac{2}{3}, 1]$, $[0, \frac{1}{4}]$, $[\frac{1}{4}, \frac{1}{2}]$, $[\frac{1}{2}, \frac{3}{4}]$, $[\frac{3}{4}, 1]$, $[0, \frac{1}{5}]$, $[\frac{1}{5}, \frac{2}{5}]$, ...

Let f_n be the characteristic function of the nth interval on this list and let f be identically zero.   If $n \geqslant m(m + 1)/2 (= 1 + 2 + \cdots + m)$, then f_n is a characteristic function of an interval whose measure is at most $1/m$.   Hence

$$\|f_n - f\|_p{}^p = \int |f_n - f|^p \, d\lambda$$

$$= \int f_n \, d\lambda \leqslant 1/m.$$

Therefore (f_n) converges in L_p to f.   However, if x is any point of $[0, 1]$, then the sequence $(f_n(x))$ has a subsequence consisting only of 1's and another subsequence consisting only of 0's.   Therefore, the sequence (f_n) *does not converge at any point of* $[0, 1]$.   (It may be observed, however, that one can select a subsequence of (f_n) which converges to f.)

CONVERGENCE IN MEASURE

Although convergence in L_p does not imply almost everywhere convergence, it does imply another type of convergence that is often of interest.

7.5 DEFINITION.   A sequence (f_n) of measurable real-valued functions is said to **converge in measure** to a measurable real-valued function f in case

$$(7.4) \qquad \lim_{n \to \infty} \mu\big(\{x \in X : |f_n(x) - f(x)| \geqslant \alpha\}\big) = 0$$

for each $\alpha > 0$.   The sequence (f_n) is said to be **Cauchy in measure** in case

$$(7.5) \qquad \lim_{m,n \to \infty} \mu\big(\{x \in X : |f_m(x) - f_n(x)| \geqslant \alpha\}\big) = 0$$

for each $\alpha > 0$.

If (f_n) converges uniformly to f, then the set

$$\{x \in X : |f_n(x) - f(x)| \geqslant \alpha\}$$

is empty for sufficiently large n.   Hence, uniform convergence implies convergence in measure.   It is not difficult to show (see Exercise 7.D) that pointwise convergence (and therefore almost everywhere convergence) need not imply convergence in measure, unless the space X has finite measure (see Theorem 7.12).   We observe, however, that convergence in L_p does imply convergence in measure.   Indeed if $E_n(\alpha) = \{x \in X : |f_n(x) - f(x)| \geqslant \alpha\}$, then

$$\int |f_n - f|^p \, d\mu \geqslant \int_{E_n(\alpha)} |f_n - f|^p \, d\mu \geqslant \alpha^p \, \mu\big(E_n(\alpha)\big).$$

Since $\alpha > 0$, it follows that $\|f_n - f\|_p \to 0$ implies that $\mu\big(E_n(\alpha)\big) \to 0$ as $n \to \infty$.

The reader can readily verify that Example 7.4 also shows that a sequence can converge in measure to a function but not converge at any point.   Despite that fact, we shall now prove a result due to F. Riesz that implies that if a sequence (f_n) converges in measure to f, then some subsequence converges almost everywhere to f.   Actually we shall prove somewhat more than that.

7.6 THEOREM.   *Let (f_n) be a sequence of measurable real-valued functions which is Cauchy in measure.   Then there is a subsequence which converges almost everywhere and in measure to a measurable real-valued function f.*

PROOF.   Select a subsequence (g_k) of (f_n) such that the set $E_k = \{x \in X : |g_{k+1}(x) - g_k(x)| \geq 2^{-k}\}$ is such that $\mu(E_k) < 2^{-k}$. Let $F_k = \bigcup_{j=k}^{\infty} E_j$ so that $F_k \in X$ and $\mu(F_k) < 2^{-(k-1)}$. If $i \geq j \geq k$ and $x \notin F_k$, then

(7.6) $|g_i(x) - g_j(x)| \leq |g_i(x) - g_{i-1}(x)| + \cdots + |g_{j+1}(x) - g_j(x)|$

$$\leq \frac{1}{2^{i-1}} + \cdots + \frac{1}{2^j} < \frac{1}{2^{j-1}}.$$

Let $F = \bigcap_{k=1}^{\infty} F_k$ so that $F \in X$ and $\mu(F) = 0$. From the argument just given it follows that (g_j) converges on $X \setminus F$. If we define f by

$$f(x) = \lim g_j(x), \qquad x \notin F,$$
$$= 0, \qquad\qquad x \in F,$$

then (g_j) converges almost everywhere to the measurable real-valued function f. Passing to the limit in (7.6) as $i \to \infty$, we infer that if $j \geq k$ and $x \notin F_k$, then

$$|f(x) - g_j(x)| \leq \frac{1}{2^{j-1}} \leq \frac{1}{2^{k-1}}.$$

This shows that the sequence (g_j) converges uniformly to f on the complement of each set F_k.

To see that (g_j) converges in measure to f, let α, ε be positive real numbers and choose k so large that $\mu(F_k) < 2^{-(k-1)} < \inf(\alpha, \varepsilon)$. If $j \geq k$, the above estimate shows that

$$\{x \in X : |f(x) - g_j(x)| \geq \alpha\} \subseteq \{x \in X : |f(x) - g_j(x)| > 2^{-(k-1)}\}$$
$$\subseteq F_k.$$

Therefore, $\mu(\{x \in X : |f(x) - g_j(x)| \geq \alpha\}) \leq \mu(F_k) < \varepsilon$ for all $j \geq k$, so that (g_j) converges in measure to f.   Q.E.D.

7.7 COROLLARY.   *Let (f_n) be a sequence of measurable real-valued functions which is Cauchy in measure. Then there is a measurable real-valued function f to which the sequence converges in measure. This limit function f is uniquely determined almost everywhere.*

PROOF. We have seen that there is a subsequence (f_{n_k}) which converges in measure to a function f. To see that the entire sequence converges in measure to f, observe that since

$$|f(x) - f_n(x)| \leqslant |f(x) - f_{n_k}(x)| + |f_{n_k}(x) - f_n(x)|,$$

it follows that

$$\{x \in X : |f(x) - f_n(x)| \geqslant \alpha\} \subseteq \left\{x \in X : |f(x) - f_{n_k}(x)| \geqslant \frac{\alpha}{2}\right\}$$
$$\bigcup \left\{x \in X : |f_{n_k}(x) - f_n(x)| \geqslant \frac{\alpha}{2}\right\}.$$

The convergence in measure of (f_n) to f follows from this relation.

Suppose that the sequence (f_n) converges in measure to both f and g. Since

$$|f(x) - g(x)| \leqslant |f(x) - f_n(x)| + |f_n(x) - g(x)|,$$

it follows that

$$\{x \in X : |f(x) - g(x)| \geqslant \alpha\} \subseteq \left\{x \in X : |f(x) - f_n(x)| \geqslant \frac{\alpha}{2}\right\}$$
$$\bigcup \left\{x \in X : |f_n(x) - g(x)| \geqslant \frac{\alpha}{2}\right\},$$

so that

$$\mu\big(\{x \in X : |f(x) - g(x)| \geqslant \alpha\}\big) = 0$$

for all $\alpha > 0$. Taking $\alpha = 1/n$, $n \in N$, we infer that $f = g$, a.e. Q.E.D.

It has been remarked that convergence in L_p implies convergence in measure. In general, convergence in measure does not imply convergence in L_p (see Exercise 7.E). However, this implication does hold when the convergence is dominated.

7.8 THEOREM. *Let (f_n) be a sequence of functions in L_p which converges in measure to f and let $g \in L_p$ be such that*

$$|f_n(x)| \leqslant g(x), \qquad \text{a.e.}$$

Then $f \in L_p$ and (f_n) converges in L_p to f.

PROOF If (f_n) does not converge in L_p to f, there exist a subsequence (g_k) of (f_n) and an $\varepsilon > 0$ such that

$$(7.7) \qquad \|g_k - f\|_p > \varepsilon \qquad \text{for} \quad k \in N.$$

Since (g_k) is a subsequence of (f_n), it follows (see Exercise 7.G) that it converges in measure to f. By Theorem 7.6 there is a subsequence (h_r) of (g_k) which converges almost everywhere and in measure to a function h. From the uniqueness part of Corollary 7.7 it follows that $h = f$ a.e. Since (h_r) converges almost everywhere to f and is dominated by g, Theorem 7.2 implies that $\|h_r - f\|_p \to 0$. However, this contradicts the relation (7.7). Q.E.D.

ALMOST UNIFORM CONVERGENCE

In the proof of Theorem 7.6 we constructed a sequence (g_j) of measurable real-valued functions which was uniformly convergent on the complement of sets which have arbitrarily small measure. At first mention this sounds equivalent to uniform convergence outside a set of zero measure, but it is not equivalent (see Exercise 7.J).

7.9 DEFINITION. A sequence (f_n) of measurable functions is said to be **almost uniformly convergent** to a measurable function f if for each $\delta > 0$ there is a set E_δ in X with $\mu(E_\delta) < \delta$ such that (f_n) converges uniformly to f on $X \setminus E_\delta$. The sequence (f_n) is said to be an **almost uniformly Cauchy sequence** if for every $\delta > 0$ there exists a set E_δ in X with $\mu(E_\delta) < \delta$ such that (f_n) is uniformly convergent on $X \setminus E_\delta$.

The reader is warned that the terminology (in addition to being unpleasant) is slightly at variance with the earlier use of the modifier "almost." It is clear that almost uniform convergence is implied by uniform convergence, but it is not hard to see that almost uniform convergence does not imply this stronger mode.

7.10 LEMMA. *Let (f_n) be an almost uniformly Cauchy sequence. Then there exists a measurable function f such that (f_n) converges almost uniformly and almost everywhere to f.*

PROOF. If $k \in N$, let $E_k \in X$ be such that $\mu(E_k) < 2^{-k}$ and (f_n) is uniformly convergent on $X \setminus E_k$. Let $F_k = \bigcup_{j=k}^{\infty} E_j$, so that $F_k \in X$ and $\mu(F_k) < 2^{-(k-1)}$. Note that (f_n) converges uniformly on $X \setminus F_k \subseteq X \setminus E_k$ and define g_k by

$$g_k(x) = \lim f_n(x), \qquad x \notin F_k,$$
$$= 0, \qquad x \in F_k.$$

We observe that the sequence (F_k) is decreasing and that if $F = \bigcap F_k$, then $F \in X$ and $\mu(F) = 0$. If $h \leqslant k$, then $g_h(x) = g_k(x)$ for all $x \notin F_h$. Therefore, the sequence (g_k) converges on all of X to a measurable limit function which we shall denote by f. If $x \notin F_k$, then $f(x) = g_k(x) = \lim f_n(x)$. It follows that (f_n) converges to f on $X \setminus F$, so that (f_n) converges to f almost everywhere on X.

To see that the convergence is almost uniform, let $\varepsilon > 0$, and let K be so large that $2^{-(K-1)} < \varepsilon$. Then $\mu(F_K) < \varepsilon$, and (f_n) converges uniformly to $g_K = f$ on $X \setminus F_K$. Q.E.D.

The next result relates convergence in measure and almost uniform convergence.

7.11 THEOREM. *If a sequence (f_n) converges almost uniformly to f, then it converges in measure. Conversely, if a sequence (h_n) converges in measure to h, then some subsequence converges almost uniformly to h.*

PROOF. Suppose that (f_n) converges almost uniformly to f, and let α and ε be positive numbers. Then there exists a set E_ε in X with $\mu(E_\varepsilon) < \varepsilon$ such that (f_n) converges to f uniformly on $X \setminus E_\varepsilon$. Therefore, if n is sufficiently large, then the set $\{x \in X : |f_n(x) - f(x)| \geqslant \alpha\}$ must be contained in E_ε. This shows that (f_n) converges in measure to f.

Conversely, suppose that (h_n) converges in measure to h. It follows from Theorem 7.6 that there is a subsequence (g_k) of (h_n) which converges in measure to a function g and the proof of Theorem 7.6 actually shows that the convergence is almost uniform. Since (g_k) converges in measure to both h and g, it follows from Corollary 7.7 that $h = g$ a.e. Therefore the subsequence (g_k) of (h_n) converges almost uniformly to h. Q.E.D.

It follows from the Theorem 7.11 that if a sequence converges in L_p, then it has a subsequence which converges almost uniformly. Conversely, it may be seen (see Exercise 7.K) that almost uniform convergence does not imply convergence in L_p in general, although it does if the convergence is dominated by a function in L_p (apply Theorem 7.8).

One of the consequences of Lemma 7.10 is that almost uniform convergence implies almost everywhere convergence. In general, the converse is false (see Exercise 7.L).   However, it is a remarkable and important fact that if the functions are real-valued and if $\mu(X) < +\infty$, then almost everywhere convergence does imply almost uniform convergence.

7.12 EGOROFF'S THEOREM. *Suppose that $\mu(X) < +\infty$ and that (f_n) is a sequence of measurable real-valued functions which converges almost everywhere on X to a measurable real-valued function f.   Then the sequence (f_n) converges almost uniformly and in measure to f.*

PROOF.   We suppose without loss of generality that (f_n) converges at every point of X to f.   If $m, n \in N$, let

$$E_n(m) = \bigcup_{k=n}^{\infty} \left\{ x \in X : |f_k(x) - f(x)| \geq \frac{1}{m} \right\},$$

so that $E_n(m)$ belongs to X and $E_{n+1}(m) \subseteq E_n(m)$.   Since $f_n(x) \to f(x)$ for all $x \in X$, it follows that

$$\bigcap_{n=1}^{\infty} E_n(m) = \emptyset.$$

Since $\mu(X) < +\infty$, we infer that $\mu\big(E_n(m)\big) \to 0$ as $n \to +\infty$.   If $\delta > 0$, let k_m be such that $\mu\big(E_{k_m}(m)\big) < \delta/2^m$ and let $E_\delta = \bigcup_{m=1}^{\infty} E_{k_m}(m)$, so that $E_\delta \in X$ and $\mu(E_\delta) < \delta$.   Observe that if $x \notin E_\delta$, then $x \notin E_{k_m}(m)$, so that

$$|f_k(x) - f(x)| < \frac{1}{m}$$

for all $k \geq k_m$.   Therefore (f_k) is uniformly convergent on the complement of E_δ.   Q.E.D.

It is convenient to have a table indicating the relations between the various modes of convergence we have been discussing.   Modifying the idea in Reference [10], we present three diagrams relating almost everywhere convergence (denoted by AE), almost uniform convergence (denoted by AU), convergence in L_p (denoted by L_p), and convergence in measure (denoted by M).   It is understood that in discussing L_p convergence, it is assumed that the functions belong to L_p.   Diagram 7.1 pertains to the case of a general measure space.   A

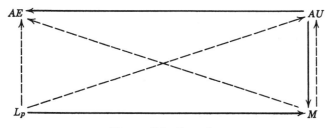

Diagram 7.1 General case

solid arrow signifies implication; a dashed arrow signifies that a subsequence converges in the indicated mode. The absence of an arrow
indicates that a counterexample can be constructed. Diagram 7.2
relates to the case of a finite measure space. In view of Egoroff's

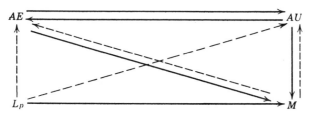

Diagram 7.2 Finite measure space

Theorem two implications are added. In Diagram 7.3, we assume
that the sequence (f_n) is dominated by a function g in L_p. Here three
implications are added.

We leave it as an exercise to verify that all the implications indicated
in these diagrams hold, and that no other ones are valid without
additional hypotheses.

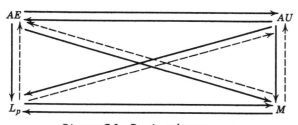

Diagram 7.3 Dominated convergence

We conclude this chapter with a set of necessary and sufficient conditions for L_p convergence. The reader will observe that the second and third conditions are automatically fulfilled when the sequence is dominated by a function in L_p.

7.13 VITALI CONVERGENCE THEOREM. *Let (f_n) be a sequence in $L_p(X, X, \mu)$, $1 \leqslant p < \infty$. Then the following three conditions are necessary and sufficient for the L_p convergence of (f_n) to f:*

(i)   *(f_n) converges to f in measure.*

(ii)   *For each $\varepsilon > 0$ there is a set $E_\varepsilon \in X$ with $\mu(E_\varepsilon) < +\infty$ such that if $F \in X$ and $F \cap E_\varepsilon = \emptyset$, .hen*

$$\int_F |f_n|^p \, d\mu < \varepsilon^p \qquad \text{for all } n \in N.$$

(iii)   *For each $\varepsilon > 0$ there is a $\delta(\varepsilon) > 0$, such that if $E \in X$ and $\mu(E) < \delta(\varepsilon)$, then*

$$\int_E |f_n|^p \, d\mu < \varepsilon^p \qquad \text{for all } n \in N.$$

PROOF.   It was seen after Definition 7.5 that L_p convergence implies convergence in measure. The fact that L_p convergence of the (f_n) implies (ii) and (iii) is not difficult and is left to the reader (see Exercises 6.R and 6.S).

We shall now show that these three conditions imply that (f_n) converges in L_p to f. If $\varepsilon > 0$, let E_ε be as in (ii) and let $F = X \setminus E_\varepsilon$ If the Minkowski Inequality is applied to $f_n - f_m = (f_n - f_m)\chi_{E_\varepsilon} + f_n \chi_F - f_m \chi_F$, we obtain

$$\|f_n - f_m\|_p \leqslant \left\{ \int_{E_\varepsilon} |f_n - f_m|^p \, d\mu \right\}^{1/p} + 2\varepsilon$$

for $n, m \in N$. Now let $\alpha = \varepsilon[\mu(E_\varepsilon)]^{-1/p}$ and let $H_{nm} = \{x \in E_\varepsilon : |f_n(x) - f_m(x)| \geqslant \alpha\}$. In view of (i), there exists a $K(\varepsilon)$ such that if $n, m \geqslant K(\varepsilon)$, then $\mu(H_{nm}) < \delta(\varepsilon)$. Another application of the Minkowski Inequality together with (iii), gives

$$\left\{ \int_{E_\varepsilon} |f_n - f_m|^p \, d\mu \right\}^{1/p} \leqslant \left\{ \int_{E_\varepsilon \setminus H_{nm}} |f_n - f_m|^p \, d\mu \right\}^{1/p}$$
$$+ \left\{ \int_{H_{nm}} |f_n|^p \, d\mu \right\}^{1/p} + \left\{ \int_{H_{nm}} |f_m|^p \, d\mu \right\}^{1/p}$$
$$\leqslant \alpha[\mu(E_\varepsilon)]^{1/p} + \varepsilon + \varepsilon = 3\varepsilon,$$

when $n, m \geqslant K(\varepsilon)$. On combining this with the earlier inequality, we infer that the sequence (f_n) is Cauchy and hence convergent in L_p. Since we already know that (f_n) is convergent in measure to f, it follows from the uniqueness in Corollary 7.7 that (f_n) converges to f in L_p.

Q.E.D.

EXERCISES

In these exercises (R, B, λ) denotes the real line with Lebesgue measure defined on the Borel subsets of R. Moreover, $1 \leqslant p < \infty$.

7.A. Let $f_n = n^{-1/p} \chi_{[0,n]}$. Show that the sequence (f_n) converges uniformly to the 0-function, but that it does not converge in $L_p(R, B, \lambda)$.

7.B. Let $f_n = n \chi_{[1/n, 2/n]}$. Show that the sequence (f_n) converges everywhere to the 0-function but that it does not converge in $L_p(R, B, \lambda)$.

7.C. Show that both of the sequences in Exercises 7.A and 7.B converge in measure to their limits.

7.D. Let $f_n = \chi_{[n, n+1]}$. Show that the sequence (f_n) converges everywhere to the 0-function, but that it does not converge in measure.

7.E. The sequence in 7.B shows that convergence in measure does not imply L_p-convergence, even for a finite measure space.

7.F. Write down a subsequence of the sequence in Example 7.4 which converges almost everywhere to the 0-function. Can you find one which converges everywhere?

7.G. If a sequence (f_n) converges in measure to a function f, then every subsequence of (f_n) converges in measure to f. More generally, if (f_n) is Cauchy in measure, then every subsequence is Cauchy in measure.

7.H. If a sequence (f_n) converges in L_p to a function f, and a subsequence of (f_n) converges in L_p to g, then $f = g$ a.e.

7.I. If (f_n) is a sequence of characteristic functions of sets in X, and if (f_n) converges to f in L_p, show that f is (almost everywhere equal to) the characteristic function of a set in X.

7.J. Show that the sequence (f_n) in Exercise 7.B has the property that if $\delta > 0$, then it is uniformly convergent on the complement of the set $[0, \delta]$.   However, show that there does not exist a set of measure zero, on the complement of which (f_n) is uniformly convergent.

7.K. Show that the sequence in Exercise 7.B converges almost uniformly but not in L_p.

7.L. Show that the sequence in Exercise 7.D converges everywhere, but not almost uniformly.

7.M. Let $f_n = n \chi_{[0, 1]}$.   Show that the hypothesis that the limit function be finite (at least almost everywhere) cannot be dropped in Egoroff's Theorem.

7.N. Show that Fatou's Lemma holds if almost everywhere convergence is replaced by convergence in measure.

7.O. Show that the Lebesgue Dominated Convergence Theorem holds if almost everywhere convergence is replaced by convergence in measure.

7.P. If $g \in L_p$ and $|f_n| \leqslant g$, show that conditions (ii) and (iii) of the Vitali Convergence Theorem 7.13 are satisfied.

7.Q. Let $(X, \mathbf{X}, \mu)$ be a finite measure space.   If f is an X-measurable function, let

$$r(f) = \int \frac{|f|}{1 + |f|} \, d\mu.$$

Show that a sequence (f_n) of X-measurable functions converges in measure to f if and only if $r(f_n - f) \to 0$.

7.R. If the sequence (f_n) of measurable functions converges almost everywhere to a measurable function f and φ is continuous on $\mathbf{R}$ to $\mathbf{R}$, then the sequence $(\varphi \circ f_n)$ converges almost everywhere to $\varphi \circ f$.   Conversely, if φ is not continuous at every point, then there exists a sequence (f_n) which converges almost everywhere to f but such that $(\varphi \circ f_n)$ does not converge almost everywhere to $\varphi \circ f$.

7.S. If φ is uniformly continuous on $\mathbf{R}$ to $\mathbf{R}$, and if (f_n) converges uniformly (respectively, almost uniformly, in measure) to f, then $(\varphi \circ f_n)$ converges uniformly (respectively, almost uniformly, in measure) to $\varphi \circ f$.   Conversely, if φ is not uniformly continuous, there exists a measure space and a sequence (f_n) converging uniformly (and hence almost uniformly and in measure) to f but such that $(\varphi \circ f_n)$ does not converge in measure (and hence not uniformly or almost uniformly) to $\varphi \circ f$.

7.T. Let (X, X, μ) be a finite measure space and let $1 \leqslant p < \infty$. Let φ be continuous on R to R and satisfy the condition: (∗) there exists $K > 0$ such that $|\varphi(t)| \leqslant K|t|$ for $|t| \geqslant K$. Show that $\varphi \circ f$ belongs to L_p for each $f \in L_p$. Conversely, if φ does not satisfy (∗), then there is a function f in L_p on a finite measure space such that $\varphi \circ f$ does not belong to L_p.

7.U. If (f_n) converges to f in L_p on a finite measure space, and if φ is continuous and satisfies condition (∗) of Exercise 7.T, then $(\varphi \circ f_n)$ converges in L_p to $\varphi \circ f$. Conversely, if condition (∗) is not satisfied, there exists a finite measure space and a sequence (f_n) which converges in L_p to f but such that $(\varphi \circ f_n)$ does not converge in L_p to $\varphi \circ f$.

7.V. Let (X, X, μ) be an arbitrary measure space. Let φ be continuous on R to R and satisfy: (∗∗) there exists $K \geqslant 0$ such that $|\varphi(t)| \leqslant K|t|$ for all $t \in R$. If $f \in L_p$, then $\varphi \circ f$ belongs to L_p. Conversely, if φ does not satisy (∗∗), there exists a measure space and a function $f \in L_p$ such that $\varphi \circ f$ does not belong to L_p.

7.W. If (f_n) converges to f in L_p on an arbitrary measure space, and if φ is continuous and satisfies (∗∗), then $(\varphi \circ f_n)$ converges to $\varphi \circ f$ in L_p. Conversely, if φ does not satisfy (∗∗), there exists a measure space and a sequence (f_n) which converges in L_p to f, but such that $(\varphi \circ f_n)$ does not converge in L_p to $\varphi \circ f$.

CHAPTER 8

Decomposition of Measures

In this chapter we shall consider the possibility of decomposing measures and charges in various ways and shall obtain some very useful results.   First we shall consider charges and show that a charge can be written as the difference of two finite measures.

We recall from Definition 3.6 that a charge on a measurable space (X, X) is a real-valued function λ defined on the σ-algebra X such that $\lambda(\emptyset) = 0$ and which is countably additive in the sense that

$$\lambda\left(\bigcup_{n=1}^{\infty} E_n \right) = \sum_{n=1}^{\infty} \lambda(E_n)$$

for any disjoint sequence (E_n) of sets in X.   The reader can easily check the proofs of Lemmas 3.3 and 3.4 to show that if (E_n) is an increasing sequence of sets in X, then

$$(8.1) \qquad \lambda\left(\bigcup_{n=1}^{\infty} E_n \right) = \lim \lambda(E_n),$$

and if (F_n) is a decreasing sequence of sets in X, then

$$(8.2) \qquad \lambda\left(\bigcap_{n=1}^{\infty} F_n \right) = \lim \lambda(F_n).$$

8.1 DEFINITION.   If λ is a charge on X, then a set P in X is said to be **positive** with respect to λ if $\lambda(E \cap P) \geqslant 0$ for any E in X.   A set N in X is said to be **negative** with respect to λ if $\lambda(E \cap N) \leqslant 0$ for any E in X.   A set M in X is said to be a **null set** for λ if $\lambda(E \cap M) = 0$ for any E in X.

It is an exercise to show that a measurable subset of a positive set is positive and that the union of two positive sets is a positive set.

8.2 HAHN DECOMPOSITION THEOREM. *If λ is a charge on X, then there exist sets P and N in X with $X = P \cup N$, $P \cap N = \emptyset$, and such that P is positive and N is negative with respect to λ.*

PROOF. The class P of all positive sets is not empty since it must contain $\emptyset$, at least. Let $\alpha = \sup \{\lambda(A) : A \in P\}$, let (A_n) be a sequence in P such that $\lim \lambda(A_n) = \alpha$, and let $P = \bigcup_{n=1}^{\infty} A_n$. Since the union of two positive sets is positive, the sequence (A_n) can be chosen to be monotone increasing, and we shall assume that this has been done. Clearly P is a positive set for λ, since

$$\lambda(E \cap P) = \lambda\left(E \cap \bigcup_{n=1}^{\infty} A_n\right) = \lambda\left(\bigcup_{n=1}^{\infty} (E \cap A_n)\right) = \lim \lambda(E \cap A_n) \geq 0.$$

Moreover, $\alpha = \lim \lambda(A_n) = \lambda(P) < \infty$.

We shall now show that the set $N = X \setminus P$ is a negative set. If not, there is a measurable subset E of N with $\lambda(E) > 0$. The set E cannot be a positive set, for then $P \cup E$ would be a positive set with $\lambda(P \cup E) > \alpha$, contrary to the definition of α. Hence E contains sets with negative charge; let n_1 be the smallest natural number such that E contains a set E_1 in X, such that $\lambda(E_1) \leq -1/n_1$. Now

$$\lambda(E \setminus E_1) = \lambda(E) - \lambda(E_1) > \lambda(E) > 0;$$

however, $E \setminus E_1$ cannot be a positive set, for then $P_1 = P \cup (E \setminus E_1)$ would be a positive set with $\lambda(P_1) > \alpha$. Therefore $E \setminus E_1$ contains sets with negative charge. Let n_2 be the smallest natural number such that $E \setminus E_1$ contains a set E_2 in X such that $\lambda(E_2) \leq -1/n_2$. As before $E \setminus (E_1 \cup E_2)$ is not a positive set, and we let n_3 be the smallest natural number such that $E \setminus (E_1 \cup E_2)$ contains a set E_3 in X such that $\lambda(E_3) \leq -1/n_3$. Repeating this argument, we obtain a disjoint sequence (E_k) of sets of X such that $\lambda(E_k) \leq -1/n_k$. Let $F = \bigcup_{k=1}^{\infty} E_k$ so that

$$\lambda(F) = \sum_{k=1}^{\infty} \lambda(E_k) \leq - \sum_{k=1}^{\infty} \frac{1}{n_k} \leq 0,$$

which shows that $1/n_k \to 0$. If G is a measurable subset of $E \setminus F$ and

$\lambda(G) < 0$, then $\lambda(G) < -1/(n_k - 1)$ for sufficiently large k, contradicting the fact that n_k is the smallest natural number such that $E \setminus (E_1 \cup \cdots \cup E_k)$ contains a set with charge less than $-1/n_k$. Hence, every measurable subset G of $E \setminus F$ must have $\lambda(G) \geqslant 0$, so that $E \setminus F$ is a positive set for λ. Since $\lambda(E \setminus F) = \lambda(E) - \lambda(F) > 0$, we infer that $P \cup (E \setminus F)$ is a positive set with charge exceeding α, which is a contradiction.

Therefore, it follows that the set $N = X \setminus P$ is a negative set for λ, and the desired decomposition of X is obtained.   Q.E.D.

A pair P, N of measurable sets satisfying the conclusions of the preceding theorem is said to form a **Hahn decomposition** of X with respect to λ.   In general, there will be no unique Hahn decomposition. In fact, if P, N is a Hahn decomposition for λ, and if M is a null set for λ, then $P \cup M$, $N \setminus M$ and $P \setminus M$, $N \cup M$ are also Hahn decompositions for λ.   This lack of uniqueness is not an important matter for most purposes, however.

8.3 LEMMA.   *If P_1, N_1 and P_2, N_2 are Hahn decompositions for λ, and E belongs to X, then*

$$\lambda(E \cap P_1) = \lambda(E \cap P_2), \qquad \lambda(E \cap N_1) = \lambda(E \cap N_2).$$

PROOF.   Since $E \cap (P_1 \setminus P_2)$ is contained in the positive set P_1 and in the negative set N_2, then $\lambda\big(E \cap (P_1 \setminus P_2)\big) = 0$ so that

$$\lambda(E \cap P_1) = \lambda(E \cap P_1 \cap P_2).$$

Similarly,

$$\lambda(E \cap P_2) = \lambda(E \cap P_1 \cap P_2),$$

from which it follows that

$$\lambda(E \cap P_1) = \lambda(E \cap P_2).$$ Q.E.D.

8.4 DEFINITION.   Let λ be a charge on X and let P, N be a Hahn decomposition for λ.   The **positive** and the **negative variations** of λ are the finite measures λ^+, λ^- defined for E in X by

(8.3) $\lambda^+(E) = \lambda(E \cap P), \qquad \lambda^-(E) = -\lambda(E \cap N).$

The **total variation** of λ is the measure $|\lambda|$ defined for E in X by

$$|\lambda|(E) = \lambda^+(E) + \lambda^-(E).$$

It is a consequence of Lemma 8.3 that the positive and negative variations are well-defined and do not depend on the Hahn decomposition. It is also clear that

(8.4) $\qquad \lambda(E) = \lambda(E \cap P) + \lambda(E \cap N) = \lambda^+(E) - \lambda^-(E).$

We shall state this result formally.

8.5 JORDAN DECOMPOSITION THEOREM. *If λ is a charge on X, it is the difference of two finite measures on X. In particular, λ is the difference of λ^+ and λ^-. Moreover, if $\lambda = \mu - \nu$ where μ, ν are finite measures on X, then*

(8.5) $\qquad \mu(E) \geqslant \lambda^+(E), \qquad \nu(E) \geqslant \lambda^-(E)$

for all E in X.

PROOF. The representation $\lambda = \lambda^+ - \lambda^-$ has already been established. Since μ and ν have nonnegative values, then

$$\lambda^+(E) = \lambda(E \cap P) = \mu(E \cap P) - \nu(E \cap P)$$
$$\leqslant \mu(E \cap P) \leqslant \mu(E).$$

Similarly, $\lambda^-(E) \leqslant \nu(E)$ for any E in X. Q.E.D.

We have seen, in Lemma 5.2, that if a function f is integrable with respect to a measure μ on X, and if λ is defined for E in X by

(8.6) $$\lambda(E) = \int_E f \, d\mu,$$

then λ is a charge. We now identify the positive and negative variations of λ.

8.6 THEOREM. *If f belongs to $L(X, X, \mu)$, and λ is defined by equation (8.6), then λ^+, λ^-, and $|\lambda|$ are given for E in X by*

$$\lambda^+(E) = \int_E f^+ \, d\mu, \quad \lambda^-(E) = \int_E f^- \, d\mu,$$
$$|\lambda|(E) = \int_E |f| \, d\mu.$$

PROOF.   Let $P_f = \{x \in X : f(x) \geqslant 0\}$ and $N_f = \{x \in X : f(x) < 0\}$. Then $X = P_f \cup N_f$ and $P_f \cap N_f = \emptyset$. If $E \in X$, then it is clear that $\lambda(E \cap P_f) \geqslant 0$ and $\lambda(E \cap N_f) \leqslant 0$. Hence P_f, N_f is a Hahn decomposition for λ. The statement now follows.   Q.E.D.

It was seen in Corollary 4.9 that if f is a nonnegative extended real-valued measurable function and μ is a measure on X, then the function λ defined by equation (8.6) is a measure on X. There is a very important converse to this which gives conditions under which one can express a measure λ as an integral with respect to μ of a non-negative extended real-valued measurable function. It was seen in Corollary 4.11 that a necessary condition for this representation is that $\lambda(E) = 0$ for any set E in X for which $\mu(E) = 0$. It turns out this condition is also sufficient in the important case where λ and μ are σ-finite.

8.7 DEFINITION.   A measure λ on X is said to be **absolutely continuous** with respect to a measure μ on X if $E \in X$ and $\mu(E) = 0$ imply that $\lambda(E) = 0$. In this case we write $\lambda \ll \mu$. A charge λ is **absolutely continuous** with respect to a charge μ in case the total variation $|\lambda|$ of λ is absolutely continuous with respect to $|\mu|$.

The following lemma is useful and adds to our intuitive understanding of absolute continuity.

8.8 LEMMA.   *Let λ and μ be finite measures on X. Then $\lambda \ll \mu$ if and only if for every $\varepsilon > 0$ there exists a $\delta(\varepsilon) > 0$ such that $E \in X$ and $\mu(E) < \delta(\varepsilon)$ imply that $\lambda(E) < \varepsilon$.*

PROOF.   If this condition is satisfied and $\mu(E) = 0$, then $\lambda(E) < \varepsilon$ for all $\varepsilon > 0$, from which it follows that $\lambda(E) = 0$.

Conversely, suppose that there exist an $\varepsilon > 0$ and sets $E_n \in X$ with $\mu(E_n) < 2^{-n}$ and $\lambda(E_n) \geqslant \varepsilon$. Let $F_n = \bigcup_{k=n}^{\infty} E_k$, so that $\mu(F_n) < 2^{-n+1}$ and $\lambda(F_n) \geqslant \varepsilon$. Since (F_n) is a decreasing sequence of measurable sets,

$$\mu\left(\bigcap_{n=1}^{\infty} F_n\right) = \lim \mu(F_n) = 0,$$

$$\lambda\left(\bigcap_{n=1}^{\infty} F_n\right) = \lim \lambda(F_n) \geqslant \varepsilon.$$

Hence λ is not absolutely continuous with respect to μ.   Q.E.D.

8.9 RADON-NIKODÝM THEOREM. *Let λ and μ be σ-finite measures defined on X and suppose that λ is absolutely continuous with respect to μ. Then there exists a function f in $M^+(X, X)$ such that*

$$(8.6) \qquad \lambda(E) = \int_E f \, d\mu, \qquad E \in X.$$

Moreover, the function f is uniquely determined μ-almost everywhere.

PROOF.   We shall first prove Theorem 8.9 under the hypothesis that $\lambda(X)$ and $\mu(X)$ are finite.

If $c > 0$, let $P(c)$, $N(c)$ be a Hahn decomposition of X for the charge $\lambda - c\mu$. If $k \in N$, consider the measurable sets

$$A_1 = N(c), \qquad A_{k+1} = N((k+1)c) \setminus \bigcup_{j=1}^{k} A_j.$$

It is clear that the sets A_k, $k \in N$, are disjoint and that

$$\bigcup_{j=1}^{k} N(jc) = \bigcup_{j=1}^{k} A_j.$$

It follows that

$$A_k = N(kc) \setminus \bigcup_{j=1}^{k-1} N(jc) = N(kc) \cap \bigcap_{j=1}^{k-1} P(jc).$$

Hence if E is a measurable subset of A_k, then $E \subseteq N(kc)$ and $E \subseteq P((k-1)c)$ so that

$$(8.7) \qquad (k-1)c\mu(E) \leqslant \lambda(E) \leqslant kc\mu(E).$$

Define B by

$$B = X \setminus \bigcup_{j=1}^{\infty} A_j = \bigcap_{j=1}^{\infty} P(jc),$$

so that $B \subseteq P(kc)$ for all $k \in N$. This implies that

$$0 \leqslant kc\mu(B) \leqslant \lambda(B) \leqslant \lambda(X) < +\infty$$

for all $k \in N$, so that $\mu(B) = 0$. Since $\lambda \ll \mu$, we infer that $\lambda(B) = 0$.

Let f_c be defined by $f_c(x) = (k-1)c$ for $x \in A_k$ and $f_c(x) = 0$ for $x \in B$. If E is an arbitrary measurable set, then E is the union of the disjoint sets $E \cap B$, $E \cap A_k$, $k \in N$, so it follows from (8.7) that

$$\int_E f_c \, d\mu \leqslant \lambda(E) \leqslant \int_E (f_c + c) \, d\mu \leqslant \int_E f_c \, d\mu + c\mu(X).$$

We employ the preceding construction for $c = 2^{-n}$, $n \in N$, to obtain a sequence of functions we now denote by f_n.   Hence

(8.8) $$\int_E f_n \, d\mu \leqslant \lambda(E) \leqslant \int_E f_n \, d\mu + 2^{-n}\mu(X).$$

for all $n \in N$.   Let $m \geqslant n$, and observe that

$$\int_E f_n \, d\mu \leqslant \lambda(E) \leqslant \int_E f_m \, d\mu + 2^{-m} \mu(X),$$

$$\int_E f_m \, d\mu \leqslant \lambda(E) \leqslant \int_E f_n \, d\mu + 2^{-n} \mu(X),$$

from which it is seen that

$$\left| \int_E (f_n - f_m) \, d\mu \right| \leqslant 2^{-n} \mu(X),$$

for all E in X.   If we let E be the sets where the integrand is positive and negative and combine, we deduce that

$$\int |f_n - f_m| \, d\mu \leqslant 2^{-n+1} \mu(X)$$

whenever $m \geqslant n$.   Thus (f_n) converges in mean and in measure to a function f.   Since the f_n belong to M^+, it is clear from Theorem 7.6 that we may require that $f \in M^+$.   Moreover,

$$\left| \int_E f_n \, d\mu - \int_E f \, d\mu \right| \leqslant \int_E |f_n - f| \, d\mu \leqslant \int |f_n - f| \, d\mu,$$

so that we conclude from (8.8) that

$$\lambda(E) = \lim \int_E f_n \, d\mu = \int_E f \, d\mu$$

for all $E \in X$.   This completes the proof of the existence assertion of the theorem in the case where both λ and μ are finite measures.

We claim that f is uniquely determined up to sets of μ-measure zero. Indeed, suppose that $f, h \in M^+$ and that

$$\lambda(E) = \int_E f \, d\mu = \int_E h \, d\mu$$

for all E in X. Let $E_1 = \{x : f(x) > h(x)\}$ and $E_2 = \{x : f(x) < h(x)\}$, and apply Corollary 4.10 to infer that $f(x) = h(x)$ μ-almost everywhere.

We shall now suppose that λ and μ are σ-finite and let (X_n) be an increasing sequence of sets in X such that

$$\lambda(X_n) < \infty, \qquad \mu(X_n) < \infty.$$

Apply the preceding argument to obtain a function h_n in M^+ which vanishes for $x \notin X_n$, such that if E is a measurable subset of X_n, then

$$\lambda(E) = \int_E h_n \, d\mu.$$

If $n \leqslant m$, then $X_n \subseteq X_m$, and it follows that

$$\int_E h_n \, d\mu = \int_E h_m \, d\mu$$

for any measurable subset E of X_n. From the uniqueness of h_n, it follows that $h_m(x) = h_n(x)$ for μ-almost all x in X_n whenever $m \geqslant n$. Let $f_n = \sup \{h_1, \ldots, h_n\}$ so that (f_n) is a monotone increasing sequence in M^+ and let $f = \lim f_n$. If $E \in X$, then

$$\lambda(E \cap X_n) = \int_E f_n \, d\mu.$$

Since $(E \cap X_n)$ is an increasing sequence of sets with union E, it follows from Lemma 3.3 and the Monotone Convergence Theorem 4.6 that

$$\lambda(E) = \lim \lambda(E \cap X_n) = \lim \int_E f_n \, d\mu$$
$$= \int_E f \, d\mu.$$

The μ-uniqueness of f is established as before. Q.E.D.

The function f whose existence we have established is often called the **Radon-Nikodým derivative** of λ with respect to μ, and is denoted by $d\lambda/d\mu$. It will be seen in the exercises to have properties closely related to the derivative. The reader should observe that this function is not necessarily integrable; in fact, f is (μ-equivalent to) an integrable function if and only if λ is a finite measure.

In intuitive terms, a measure λ is absolutely continuous with respect to a measure μ in case sets which have small μ-measure also have small λ-measure. At the opposite extreme, there is the notion of singular measures, which we now introduce.

8.10 DEFINITION. Two measures λ, μ on X are said to be **mutually singular** if there are disjoint sets A, B in X such that $X = A \cup B$ and $\lambda(A) = \mu(B) = 0$. In this case we write $\lambda \perp \mu$.

Although the relation of singularity is symmetric in λ and μ, we shall sometimes say that λ is **singular with respect to** μ.

8.11 LEBESGUE DECOMPOSITION THEOREM. *Let λ and μ be σ-finite measures defined on a σ-algebra X. Then there exists a measure λ_1 which is singular with respect to μ and a measure λ_2 which is absolutely continuous with respect to μ such that $\lambda = \lambda_1 + \lambda_2$. Moreover, the measures λ_1 and λ_2 are unique.*

PROOF. Let $\nu = \lambda + \mu$ so that ν is a σ-finite measure. Since λ and μ are both absolutely continuous with respect to ν, the Radon-Nikodým Theorem implies that there exist functions f, g in $M^+(X, X)$ such that

$$\lambda(E) = \int_E f \, d\nu, \qquad \mu(E) = \int_E g \, d\nu$$

for all E in X. Let $A = \{x : g(x) = 0\}$, and let $B = \{x : g(x) > 0\}$, so that $A \cap B = \emptyset$, and $X = A \cup B$.

Define λ_1 and λ_2 for E in X by

$$\lambda_1(E) = \lambda(E \cap A), \qquad \lambda_2(E) = \lambda(E \cap B).$$

Since $\mu(A) = 0$, it follows that $\lambda_1 \perp \mu$. To see that $\lambda_2 \ll \mu$, observe that if $\mu(E) = 0$, then

$$\int_E g \, d\nu = 0,$$

so that $g(x) = 0$ for ν-almost all x in E. Hence $\nu(E \cap B) = 0$; since $\lambda \ll \nu$,

$$\lambda_2(E) = \lambda(E \cap B) = 0.$$

Clearly $\lambda = \lambda_1 + \lambda_2$, so the existence of this decomposition is affirmed.

To establish the uniqueness of the decomposition, use the observation that if α is a measure such that $\alpha \ll \mu$, and $\alpha \perp \mu$, then $\alpha = 0$.   Q.E.D.

RIESZ REPRESENTATION THEOREM

As another application of the Radon–Nikodým Theorem, we shall present theorems concerning the representation of bounded linear functionals on the spaces L_p, $1 \leqslant p < \infty$.

8.12 DEFINITION. A **linear functional** on $L_p = L_p(X, \mathbf{X}, \mu)$ is a mapping G of L_p into $\mathbf{R}$ such that

$$G(af + bg) = aG(f) + bG(g)$$

for all a, b in $\mathbf{R}$ and f, g in L_p. The linear functional G is **bounded** if there exists a constant M such that

$$|G(f)| \leqslant M\|f\|_p$$

for all f in L_p. In this case, the **bound** or the **norm** of G is defined to be

$$(8.9) \qquad \|G\| = \sup\{|G(f)| : f \in L_p, \|f\|_p \leqslant 1\}.$$

It is a consequence of the linearity of the integral and Hölder's Inequality that if $g \in L_q$ (where $q = \infty$ when $p = 1$ and $q = p/(p - 1)$ otherwise) and if we define G on L_p by

$$(8.10) \qquad G(f) = \int fg \, d\mu,$$

then G is a linear functional with norm at most equal to $\|g\|_q$ (and it is an exercise to prove that $\|G\| = \|g\|_q$). The Riesz Theorem yields a converse to this observation.

Before we prove this theorem it is convenient to observe that any bounded linear functional on L_p can be written as the difference of two positive linear functionals (that is, functionals G such that $G(f) \geqslant 0$ for all $f \in L_p$ for which $f \geqslant 0$).

8.13 LEMMA. *Let G be a bounded linear functional on L_p. Then there exist two positive bounded linear functionals G^+, G^- such that $G(f) = G^+(f) - G^-(f)$ for all $f \in L_p$.*

PROOF. If $f \geqslant 0$ define $G^+(f) = \sup\{G(g) : g \in L_p, 0 \leqslant g \leqslant f\}$. It is clear that $G^+(cf) = c\, G^+(f)$ for $c \geqslant 0$ and $f \geqslant 0$. If $0 \leqslant g_j \leqslant f_j$, then

$$G(g_1) + G(g_2) = G(g_1 + g_2) \leqslant G^+(f_1 + f_2).$$

Taking the suprema over all such g_j in L_p we obtain $G^+(f_1) + G^+(f_2) \leqslant G^+(f_1 + f_2)$. Conversely, if $0 \leqslant h \leqslant f_1 + f_2$, let $g_1 = \sup (h - f_2, 0)$ and $g_2 = \inf (h, f_2)$. It follows that $g_1 + g_2 = h$ and that $0 \leqslant g_j \leqslant f_j$. Therefore $G(h) = G(g_1) + G(g_2) \leqslant G^+(f_1) + G^+(f_2)$; since this holds for all such $h \in L_p$, we infer that

$$G^+(f_1 + f_2) = G^+(f_1) + G^+(f_2)$$

for all f_j in L_p such that $f_j \geqslant 0$.

If f is an arbitrary element of L_p, define

$$G^+(f) = G^+(f^+) - G^+(f^-).$$

It is an elementary exercise to show that G^+ is a bounded linear functional on L_p. Further, we define G^- for $f \in L_p$ by

$$G^-(f) = G^+(f) - G(f),$$

so that G^- is evidently a bounded linear functional. From the definition of G^+ it is readily seen that G^- is a positive linear functional, and it is obvious that $G = G^+ - G^-$.   Q.E.D.

8.14 Riesz Representation Theorem. *If $(X, \mathbf{X}, \mu)$ is a σ-finite measure space and G is a bounded linear functional on $L_1(X, \mathbf{X}, \mu)$, then there exists a g in $L_\infty (X, \mathbf{X}, \mu)$ such that equation (8.10) holds for all f in L_1. Moreover, $\|G\| = \|g\|_\infty$ and $g \geqslant 0$ if G is a positive linear functional.*

proof.   We shall first suppose that $\mu(X) < \infty$ and that G is positive. Define λ on $\mathbf{X}$ to $\mathbf{R}$ by $\lambda(E) = G(\chi_E)$; clearly $\lambda(\emptyset) = 0$. If (E_n) is an increasing sequence in $\mathbf{X}$ and $E = \bigcup E_n$, then (χ_{E_n}) converges pointwise to χ_E. Since $\mu(X) < \infty$, it follows from Corollary 7.3 that this sequence converges in L_1 to χ_E. Since

$$0 \leqslant \lambda(E) - \lambda(E_n) = G(\chi_E) - G(\chi_{E_n})$$
$$= G(\chi_E - \chi_{E_n}) \leqslant \|G\| \, \|\chi_E - \chi_{E_n}\|_1,$$

it follows that λ is a measure. Moreover, if $M \in \mathbf{X}$ and $\mu(M) = 0$, then $\lambda(M) = 0$, so that $\lambda \ll \mu$.

On applying the Radon-Nikodým Theorem we obtain a nonnegative measurable function on X to $\mathbf{R}$ such that

$$G(\chi_E) = \lambda(E) = \int \chi_E \, g \, d\mu$$

for all $E \in X$. It follows by linearity that

$$G(\varphi) = \int \varphi g \, d\mu$$

for all X-measurable simple functions φ.

If f is a nonnegative function in L_1, let (φ_n) be a monotone increasing sequence of simple functions converging almost everywhere and in L_1 to f. From the boundedness of G it is seen that $G(f) = \lim G(\varphi_n)$. Moreover, it follows from the Monotone Convergence Theorem that

$$G(f) = \lim_{n} \int \varphi_n g \, d\mu = \int fg \, d\mu.$$

This relation holds for arbitrary $f \in L_1$ by linearity.

We now turn to the σ-finite case. If $X = \bigcup F_n$, where (F_n) is an increasing sequence of sets in X with finite measure, the preceding argument yields the existence of nonnegative functions g_n such that

$$G(f \chi_{F_n}) = \int f \chi_{F_n} g_n \, d\mu$$

for all f in L_1. If $m \leqslant n$ it is readily seen that $g_m(x) = g_n(x)$ for almost all x in F_m. In this way we obtain a function g which represents G.

If G is an arbitrary bounded linear functional on L_1, Lemma 8.13 shows that we can write $G = G^+ - G^-$, where G^+ and G^- are bounded positive linear functionals. If we apply the preceding considerations to G^+ and G^-, we obtain nonnegative measurable functions g^+, g^- which represent G^+, G^-. If we set $g = g^+ - g^-$, we obtain the representation

(8.10)
$$G(f) = \int fg \, d\mu$$

for all $f \in L_1$. It will be left as an exercise to show that $\|G\| = \|g\|_\infty$.

Q.E.D.

8.15 RIESZ REPRESENTATION THEOREM. *If (X, X, μ) is an arbitrary measure space and G is a bounded linear functional on $L_p(X, X, \mu)$, $1 < p < \infty$, then there exists a g in $L_q(X, X, \mu)$, where $q = p/(p - 1)$, such that equation (8.10) holds for all f in L_p. Moreover, $\|G\| = \|g\|_q$.*

PROOF. If $\mu(X) < \infty$, the proof of the preceding theorem requires only minor changes to show that there exists a g in L_q with $\|G\| = \|g\|_q$ and such that

$$G(f) = \int fg \, d\mu$$

for all f in L_p. In addition, the procedure used before applies to extend the result to the case where (X, X, μ) is σ-finite.

We now complete the proof by observing that a bounded linear functional "vanishes off of a σ-finite set." More precisely, let (f_n) be a sequence in L_p such that $\|f_n\| = 1$ and

$$G(f_n) \geqslant \|G\| \left(1 - \frac{1}{n}\right).$$

There exists a σ-finite set X_0 in X outside of which all the f_n vanish. Let $E \in X$ with $E \cap X_0 = \emptyset$, then $\|f_n \pm t\chi_E\|_p \leqslant (1 + t^p \mu(E))^{1/p}$ for $t \geqslant 0$, when $\mu(E) < \infty$. Moreover, since

$$G(f_n) + G(\pm t\chi_E) \leqslant |G(f_n \pm t \chi_E)|,$$

it follows that

$$|G(t\chi_E)| \leqslant \|G\| \left\{ \left(1 + t^p \mu(E)\right)^{1/p} - \left(1 - \frac{1}{n}\right) \right\}.$$

for all n in N. First let $n \to \infty$, and then divide by $t > 0$, to get

$$|G(\chi_E)| \leqslant \|G\| \frac{(1 + t^p \mu(E))^{1/p} - 1}{t}.$$

If we apply L'Hospital's Rule as $t \to 0+$, we infer that $G(\chi_E) = 0$, for any $E \in X$, $\mu(E) < \infty$, outside of the σ-finite set X_0. Therefore if f is any function in L_p such that $X_0 \cap \{x \in X : f(x) \neq 0\} = \emptyset$, it follows that $G(f) = 0$.

Hence we can apply the preceding argument to obtain a function g on X_0 which represents G, and extend g to all of X by requiring that it vanish on the complement of X_0. In this way we obtain the desired function.   Q.E.D.

EXERCISES

8.A. If P is a positive set with respect to a charge λ, and if $E \in X$ and $E \subseteq P$, then E is positive with respect to λ.

8.B. If P_1 and P_2 are positive sets for a charge λ, then $P_1 \cup P_2$ is positive for λ.

8.C. A set M in X is a null set for a charge λ if and only if $|\lambda|(M) = 0$.

8.D. If λ is a charge on X, then the values of λ are bounded and

$$\lambda^+(E) = \sup\{\lambda(F) : F \subseteq E, F \in X\},$$
$$\lambda^-(E) = -\inf\{\lambda(F) : F \subseteq E, F \in X\}.$$

8.E. Let μ_1, μ_2, and μ_3 be measures on (X, X). Show that $\mu_1 \ll \mu_1$ and that $\mu_1 \ll \mu_2$ and $\mu_2 \ll \mu_3$ imply that $\mu_1 \ll \mu_3$. Give an example to show that $\mu_1 \ll \mu_2$ does not imply that $\mu_2 \ll \mu_1$.

8.F. If (μ_n) is a sequence of measures on (X, X) with $\mu_n(X) \leqslant 1$, let λ be defined for E in X by

$$\lambda(E) = \sum_{n=1}^{\infty} 2^{-n} \mu_n(E).$$

Show that λ is a measure and that $\mu_n \ll \lambda$ for all n.

8.G. Let λ be a charge and let μ be a measure on (X, X). If $\lambda \ll \mu$, then λ^+, λ^-, and $|\lambda|$ are absolutely continuous with respect to μ.

8.H. Show that Lemma 8.8 is true even if μ is allowed to be an infinite measure. However, it may fail if λ is an infinite measure. [*Hint:* Let λ be the counting measure on N, and let

$$\mu(E) = \sum_{n \in E} 2^{-n}.]$$

8.I. Let μ be defined as in Exercise 8.H and if $E \subseteq N$, let λ be defined by
$$\lambda(E) = 0, \quad \text{if} \quad E = \emptyset;$$
$$= +\infty, \quad \text{if} \quad E \neq \emptyset.$$

Show that μ is a finite measure on the σ-algebra X of all subsets of N, and that λ is an infinite measure on X. Moreover, $\lambda \ll \mu$ and $\mu \ll \lambda$.

8.J. If λ and μ are σ-finite and $\lambda \ll \mu$, then the function f in the Radon-Nikodým Theorem can be taken to be finite-valued on X.

8.K. Let μ be a finite measure, let $\lambda \ll \mu$, and let P_n, N_n be a Hahn decomposition for $\lambda - n\mu$. Let $P = \bigcap P_n$, $N = \bigcup N_n$. Show that N is σ-finite for λ and that if $E \subseteq P$, $E \in X$, then either $\lambda(E) = 0$ or $\lambda(E) = +\infty$.

8.L. Use Exercise 8.K to extend the Radon-Nikodým Theorem to the case where μ is σ-finite and λ is an arbitrary measure with $\lambda \ll \mu$. Here f is not necessarily finite-valued.

8.M. (a) Let X be an uncountable set and X be the family of all subsets E of X such that either E or $X \setminus E$ is countable. Let $\mu(E)$ equal the number of elements in E if E is finite and equal $+\infty$ otherwise, and let $\lambda(E) = 0$ if E is countable and equal $+\infty$ if E is uncountable. Then $\lambda \ll \mu$, but the Radon-Nikodým Theorem fails.

(b) Let $X = [0, 1]$ and let X be the Borel subsets of X. If μ is the counting measure on X and λ is Lebesgue measure on X, then λ is a finite measure and $\lambda \ll \mu$, but the Radon-Nikodým Theorem fails.

8.N. Let λ, μ be σ-finite measures on (X, X), let $\lambda \ll \mu$, and let $f = d\lambda/d\mu$. If g belongs to $M^+(X, X)$, then

$$\int g \, d\lambda = \int gf \, d\mu.$$

(*Hint:* First consider simple functions and apply the Monotone Convergence Theorem.)

8.O. Let λ, μ, ν be σ-finite measures on (X, X). Use Exercise 8.N to show that if $\nu \ll \lambda$ and $\lambda \ll \mu$, then

$$\frac{d\nu}{d\mu} = \frac{d\nu}{d\lambda} \frac{d\lambda}{d\mu}, \quad \mu\text{-almost everywhere.}$$

Also, if $\lambda_j \ll \mu$ for $j = 1, 2$, then

$$\frac{d}{d\mu}(\lambda_1 + \lambda_2) = \frac{d\lambda_1}{d\mu} + \frac{d\lambda_2}{d\mu}, \quad \mu\text{-almost everywhere.}$$

8.P. If λ and μ are σ-finite, $\lambda \ll \mu$, and $\mu \ll \lambda$, then

$$\frac{d\lambda}{d\mu} = \frac{1}{d\mu/d\lambda}, \quad \text{almost everywhere.}$$

8.Q. If λ and μ are measures, with $\lambda \ll \mu$ and $\lambda \perp \mu$, then $\lambda = 0$.

8.R. If λ is a charge and μ is a measure, then $|\lambda| \perp \mu$ implies that λ^+ and λ^- are singular with respect to μ.

8.S. The collection of all charges on $(X, \mathbf{X})$ is a Banach space under the vector operations

$$(c\mu)(E) = c\mu(E), \quad (\lambda + \mu)(E) = \lambda(E) + \mu(E)$$

and the norm $\|\mu\| = |\mu|(X)$.

8.T. Suppose g satisfies equation (8.10) for all f in L_1 and that $c > 1$. Let $E_c = \{x : |g(x)| \geqslant c\|G\|\}$, and define $f_c(x)$ to be ± 1 when $\pm g(x) \geqslant c\|G\|$ and to be 0 when $x \notin E_c$. Then

$$c\|G\|\mu(E_c) \leqslant G(f_c) \leqslant \|G\|\mu(E_c),$$

which is a contradiction unless $\mu(E_c) = 0$. Infer that $|g(x)| \leqslant \|G\|$ for μ-almost all x.

8.U. If g satisfies (8.10) for all $f \in L_p$, show that $g \in L_q$ and that $\|G\| = \|g\|_q$.

8.V. The Riesz Representation Theorem for $p = 2$ can be proved by some elementary Hilbert space geometry (see [5], pp. 249–50). We now show that this result can be used to prove the Radon-Nikodým Theorem. We shall limit our attention to finite measures λ, μ with $\lambda \ll \mu$. Let $\nu = \lambda + \mu$ and show that

$$G(f) = \int f \, d\lambda$$

defines a positive linear functional on $L_2(X, \mathbf{X}, \nu)$ with positive norm. If $g \in L_2(X, \mathbf{X}, \nu)$ is such that

$$G(f) = \int fg \, d\nu, \quad f \in L_2(X, \mathbf{X}, \nu),$$

then we see by taking $f = \chi_E$, $E \in \mathbf{X}$, that $0 \leqslant g(x) \leqslant 1$ for ν-almost all x. Moreover, $\mu\{x : g(x) = 1\} = 0$. Since $\nu = \lambda + \mu$, we have

$$\int h (1 - g) \, d\lambda = \int h g \, d\mu$$

for all nonnegative $h \in L_2(X, \mathbf{X}, \nu)$ and hence for all nonnegative measurable h. Now take $h = \chi_E/(1 - g)$ to infer that

$$\lambda(E) = \int_E \left(\frac{g}{1 - g}\right) d\mu.$$

CHAPTER 9

Generation of Measures

In the preceding chapters we have given a few examples of measures, but they are of a rather special form, and it is time to demonstrate how measures can be constructed. In particular, we wish to show how to construct Lebesgue measure on the real line R from the length of an interval.

It is natural to define the **length** of the half-open interval $(a, b]$ to be the real number $b - a$ and the length of the sets $(-\infty, b] = \{x \in R : x \leqslant b\}$, and $(a, +\infty) = \{x \in R : a < x\}$, and $(-\infty, +\infty)$ to be the extended real number $+\infty$. We define the length of the union of a finite number of disjoint sets of these forms to be the sum of the corresponding lengths. Thus, the length of

$$\bigcup_{j=1}^{n} (a_j, b_j] \quad \text{is} \quad \sum_{j=1}^{n} (b_j - a_j)$$

provided the intervals do not intersect.

At first glance one might think that we have defined a measure on the family F of all sets which are finite unions of sets of the form

(9.1) $\qquad (a, b], \quad (-\infty, b], \quad (a, +\infty), \quad (-\infty, +\infty).$

However, this is not the case since the countable union of sets in F is not necessarily in F, so that F is not a σ-algebra in the sense of Definition 2.1.

9.1 DEFINITION. A family A of subsets of a set X is said to be an **algebra** or a **field** in case:

(i) $\emptyset$, X belong to A.

(ii) If E belongs to A, then its complement $X \setminus E$ also belongs to A.

(iii) If $E_1, \ldots, E_n$ belong to A, then their union $\bigcup_{j=1}^{n} E_j$ also belongs to A.

It is convenient to define the notion of a measure on an algebra. In doing so, we require the set function to be countably additive over sequences whose union belongs to the algebra.

9.2 DEFINITION. If A is an algebra of subsets of a set X, then a **measure** on A is an extended real-valued function μ defined on A such that (i) $\mu(\emptyset) = 0$, (ii) $\mu(E) \geqslant 0$ for all $E \in A$, and (iii) if (E_n) is any disjoint sequence of sets in A such that $\bigcup_{n=1}^{\infty} E_n$ belongs to A, then

$$\mu\left(\bigcup_{n=1}^{\infty} E_n\right) = \sum_{n=1}^{\infty} \mu(E_n).$$

It seems reasonably clear, but not entirely obvious, that length gives a measure. We now prove this fact.

9.3 LEMMA. *The collection* F *of all finite unions of sets of the form* (9.1) *is an algebra of subsets of* R *and length is a measure on* F.

PROOF. It is readily seen that F is an algebra. If l denotes the length function, then conditions 9.2(i) and (ii) are trivial. To prove (iii) it is enough to show that if one of the sets of the form (9.1) is the union of a countable collection of sets of this form, then the length adds up correctly. We shall treat an interval of the form $(a, b]$, leaving the other possibilities as exercises. Suppose, then, that

$$(9.2) \qquad\qquad (a, b] = \bigcup_{j=1}^{\infty} (a_j, b_j],$$

where the intervals $(a_j, b_j]$ are disjoint. Let $(a_1, b_1], \ldots, (a_n, b_n]$ be any finite collection of such intervals and suppose that

$$a \leqslant a_1 < b_1 \leqslant a_2 < \cdots < b_{n-1} \leqslant a_n < b_n \leqslant b.$$

(This may require a renumbering of the indices, but it can always be arranged.) Now

$$\sum_{j=1}^{n} l\big((a_j, b_j]\big) = \sum_{j=1}^{n} (b_j - a_j)$$
$$\leqslant b_n - a_1 \leqslant b - a = l\big((a, b]\big).$$

Since n is arbitrary, we infer that

$$(9.3) \qquad \sum_{j=1}^{\infty} l((a_j, b_j]) \leqslant ((a, b]).$$

Conversely, let $\varepsilon > 0$ be arbitrary, and let (ε_j) be a sequence of positive numbers with $\sum \varepsilon_j < \varepsilon/2$. Now consider the intervals

$$I_j = (a_j - \varepsilon_j, b_j + \varepsilon_j), \qquad j \in \mathbf{N}.$$

From (9.2) it follows that the open sets $\{I_j : j \in \mathbf{N}\}$ form a covering of the compact interval $[a, b]$.   Hence, this interval is covered by a finite number of the intervals, say by $I_1, I_2, \ldots, I_m$.   By renumbering and discarding some extra intervals we may assume that

$$a_1 - \varepsilon_1 < a, \qquad b < b_m + \varepsilon_m,$$
$$a_j - \varepsilon_j < b_{j-1} + \varepsilon_{j-1}, \qquad j = 2, \ldots, m.$$

It follows from these inequalities that

$$b - a \leqslant (b_m + \varepsilon_m) - (a_1 - \varepsilon_1) \leqslant \sum_{j=1}^{m}[(b_j + \varepsilon_j) - (a_j - \varepsilon_j)]$$
$$\leqslant \sum_{j=1}^{m}(b_j - a_j) + \varepsilon \leqslant \sum_{j=1}^{\infty}(b_j - a_j) + \varepsilon.$$

Since $\varepsilon > 0$ is arbitrary, it follows that $l((a, b]) \leqslant \sum_{j=1}^{\infty} l((a_j, b_j])$. Combining this inequality with (9.3), we conclude that the length function l is countably additive on $\mathbf{F}$.   Q.E.D.

THE EXTENSION OF MEASURES

Now that we have given a significant example of a measure defined on an algebra of sets, we return to the general situation. We shall show that if A is any algebra of subsets of a set X and if μ is a measure defined on A, then there exists a σ-algebra A^* containing A and a measure μ^* defined on A^* such that $\mu^*(E) = \mu(E)$ for E in A.   In other words, the measure μ can be extended to a measure on a σ-algebra A^* of subsets of X which contains A. The procedure that we employ is the following: we shall use μ to obtain a function defined for *all* subsets of X, and then pick out a collection of sets on which a certain additivity property holds.

9.4 DEFINITION. If B is an arbitrary subset of X, we define

$$(9.4) \qquad \mu^*(B) = \inf \sum_{j=1}^{\infty} \mu(E_j),$$

where the infimum is extended over all sequences (E_j) of sets in A such that

$$(9.5) \qquad B \subseteq \bigcup_{j=1}^{\infty} E_j.$$

It should be remarked that the function μ^* just defined is usually called the **outer measure** generated by μ. Although this terminology is unfortunate because μ^* is not generally a measure, μ^* does have a few properties reminiscent of a measure.

9.5 LEMMA. *The function μ^* of Definition 9.4 satisfies the following:*

(a) $\mu^*(\emptyset) = 0$.
(b) $\mu^*(B) \geqslant 0$, *for $B \subseteq X$.*
(c) *If $A \subseteq B$, then $\mu^*(A) \leqslant \mu^*(B)$.*
(d) *If $B \in A$, then $\mu^*(B) = \mu(B)$.*
(e) *If (B_n) is a sequence of subsets of X, then*

$$\mu^* \left(\bigcup_{n=1}^{\infty} B_n \right) \leqslant \sum_{n=1}^{\infty} \mu^*(B_n).$$

PROOF. Statements (a), (b), and (c) are immediate consequences of the Definition 9.4.

(d) Since $\{B, \emptyset, \emptyset, \ldots\}$ is a countable collection of sets in A whose union contains B, it follows that

$$\mu^*(B) \leqslant \mu(B) + 0 + 0 + \cdots = \mu(B).$$

Conversely, if (E_n) is any sequence from A with $B \subseteq \bigcup E_n$, then

$B = \bigcup (B \cap E_n)$.   Since μ is a measure on A, then

$$\mu(B) \leqslant \sum_{n=1}^{\infty} \mu(B \cap E_n) \leqslant \sum_{n=1}^{\infty} \mu(E_n),$$

from which it follows that $\mu(B) \leqslant \mu^*(B)$.

To establish (e), let $\varepsilon > 0$ be arbitrary and for each n choose a sequence (E_{nk}) of sets in A such that

$$B_n \subseteq \bigcup_{k=1}^{\infty} E_{nk} \quad \text{and} \quad \sum_{k=1}^{\infty} \mu(E_{nk}) \leqslant \mu^*(B_n) + \frac{\varepsilon}{2^n}.$$

Since $\{E_{nk} : n, k \in N\}$ is a countable collection from A whose union contains $\bigcup B_n$, it follows from the definition of μ^* that

$$\mu^*\left(\bigcup_{n=1}^{\infty} B_n \right) \leqslant \sum_{n=1}^{\infty} \sum_{k=1}^{\infty} \mu(E_{nk}) \leqslant \sum_{n=1}^{\infty} \mu^*(B_n) + \varepsilon.$$

Since ε is arbitrary, the desired inequality is obtained.   Q.E.D.

Property (e) of Lemma 9.5 is referred to by saying that μ^* is **countably subadditive**.

Although μ^* has the advantage that it is defined for arbitrary subsets of X, it has the defect that it is not necessarily countably (or even finitely) additive.   We are willing to restrict μ^* to a smaller σ-algebra provided we can find one containing A and over which μ^* has the property of countable additivity.   There is a remarkable condition due to Carathéodory which provides the desired restriction of the domain of μ^*.

9.6 DEFINITION.   A subset E of X is said to be μ^*-**measurable** if

(9.6) $$\mu^*(A) = \mu^*(A \cap E) + \mu^*(A \setminus E)$$

for all subsets A of X.   The collection of all μ^*-measurable sets is denoted by A^*.

Condition (9.6) indicates an additivity property on μ^*.   In loose terms, a set E is μ^*-measurable in case it and its complement are sufficiently separated that they divide an arbitrary set A additively.

9.7 CARATHÉODORY EXTENSION THEOREM. *The collection A^* of all μ^*-measurable sets is a σ-algebra containing A. Moreover, if (E_n) is a disjoint sequence in A^*, then*

$$(9.7) \qquad \mu^*\left(\bigcup_{n=1}^{\infty} E_n\right) = \sum_{n=1}^{\infty} \mu^*(E_n).$$

PROOF. It is clear that $\emptyset$ and X are μ^*-measurable, and that if $E \in A^*$, then its complement $X \setminus E$ belongs to A^*.

Next we shall show that A^* is closed under intersections. Indeed, suppose that E and F are μ^*-measurable. Then for any $A \subseteq X$ and $E \in A^*$, we have

$$\mu^*(A \cap F) = \mu^*(A \cap F \cap E) + \mu^*\big((A \cap F) \setminus E\big)$$

Since $F \in A^*$, then

$$\mu^*(A) = \mu^*(A \cap F) + \mu^*(A \setminus F).$$

Let $B = A \setminus (E \cap F)$, then it is readily seen that $B \cap F = (A \cap F) \setminus E$ and $B \setminus F = A \setminus F$; since $F \in A^*$ it follows that

$$\mu^*\big(A \setminus (E \cap F)\big) = \mu^*\big((A \cap F) \setminus E\big) + \mu^*(A \setminus F).$$

Combining these three relations, we obtain

$$\mu^*(A) = \mu^*(A \cap E \cap F) + \mu^*\big(A \setminus (E \cap F)\big),$$

which shows that $E \cap F$ belongs to A^*. Since A^* is closed under intersection and complementation, it follows that A^* is an algebra.

Suppose that $E, F \in A^*$ and that $E \cap F = \emptyset$. If we take A to be $A \cap (E \cup F)$ in (9.6), we obtain

$$\mu^*\big(A \cap (E \cup F)\big) = \mu^*(A \cap E) + \mu^*(A \cap F).$$

For $A = X$, this relation implies that μ^* is additive on A^*.

We shall now show that A^* is a σ-algebra and that μ^* is countably additive on A^*. Let (E_k) be a disjoint sequence in A^* and let $E = \bigcup E_k$. From the preceding paragraph, we know that $F_n = \bigcup_{k=1}^{n} E_k$ belongs to A^*, and that if A is any subset of X, then

$$\mu^*(A) = \mu^*(A \cap F_n) + \mu^*(A \setminus F_n) = \sum_{k=1}^{n} \mu^*(A \cap E_k) + \mu^*(A \setminus F_n).$$

Since $F_n \subseteq E$, then $A \setminus E \subseteq A \setminus F_n$ and letting $n \to \infty$ the above relations yields

$$\sum_{k=1}^{\infty} \mu^*(A \cap E_k) + \mu^*(A \setminus E) \leqslant \mu^*(A).$$

On the other hand, it follows from Lemma 9.5(e) that

$$\mu^*(A \cap E) \leqslant \sum_{k=1}^{\infty} \mu^*(A \cap E_k),$$
$$\mu^*(A) \leqslant \mu^*(A \cap E) + \mu^*(A \setminus E).$$

On combining the last three inequalities we infer that

$$\mu^*(A) = \mu^*(A \cap E) + \mu^*(A \setminus E) = \sum_{k=1}^{\infty} \mu^*(A \cap E_k) + \mu^*(A \setminus E).$$

In particular, this shows that $E = \bigcup_{k=1}^{\infty} E_k$ is μ^*-measurable. On taking $A = E$, we obtain (9.7).

It remains to prove that $\mathbf{A} \subseteq \mathbf{A}^*$. It was proved in Lemma 9.5(d) that if $E \in \mathbf{A}$, then $\mu^*(E) = \mu(E)$, but we need to show that E is μ^*-measurable. Let A be an arbitrary subset of X; it follows from Lemma 9.5(e) that

$$\mu^*(A) \leqslant \mu^*(A \cap E) + \mu^*(A \setminus E).$$

To establish the opposite inequality, let $\varepsilon > 0$ be arbitrary and let (F_n) be a sequence in $\mathbf{A}$ such that $A \subseteq \bigcup F_n$ and

$$\sum_{n=1}^{\infty} \mu(F_n) \leqslant \mu^*(A) + \varepsilon.$$

Since $A \cap E \subseteq \bigcup (F_n \cap E)$ and $A \setminus E \subseteq \bigcup (F_n \setminus E)$, it follows from Lemma 9.5(e) that

$$\mu^*(A \cap E) \leqslant \sum_{n=1}^{\infty} \mu(F_n \cap E), \qquad \mu^*(A \setminus E) \leqslant \sum_{n=1}^{\infty} \mu(F_n \setminus E).$$

Hence we have

$$\mu^*(A \cap E) + \mu^*(A \setminus E) \leqslant \sum_{n=1}^{\infty} \{\mu(F_n \cap E) + \mu(F_n \setminus E)\}$$
$$= \sum_{n=1}^{\infty} \mu(F_n) \leqslant \mu^*(A) + \varepsilon.$$

Since ε is arbitrary, the desired inequality is established and the set E belongs to A^*. Q.E.D.

The Carathéodory Extension Theorem shows that a measure μ on an algebra A can always be extended to a measure μ^* on a σ-algebra A^* containing A. The σ-algebra A^* obtained in this way is automatically **complete** in the sense that if $E \in A^*$ with $\mu^*(E) = 0$, and if $B \subseteq E$, then $B \in A^*$ and $\mu^*(B) = 0$. To prove this, let A be an arbitrary subset of X and employ Lemma 9.5(c) to observe that

$$\mu^*(A) = \mu^*(E) + \mu^*(A) \geqslant \mu^*(A \cap B) + \mu^*(A \setminus B);$$

and, as before, the inequality

$$\mu^*(A) \leqslant \mu^*(A \cap B) + \mu^*(A \setminus B)$$

follows from Lemma 9.5(e). Hence B is μ^*-measurable and

$$0 \leqslant \mu^*(B) \leqslant \mu^*(E) \leqslant 0.$$

We shall now show that in the case that μ is a σ-finite measure, it has a unique extension to a measure on A^*.

9.8 HAHN EXTENSION THEOREM. *Suppose that μ is a σ-finite measure on an algebra A. Then there exists a unique extension of μ to a measure on A^*.*

PROOF. The fact that μ^* gives a measure on A^* was proved in Theorem 9.7 even without the σ-finiteness assumption. To establish the uniqueness, let ν be a measure on A^* which agrees with μ on A.

First suppose that μ and therefore both μ^* and ν are finite measures. Let E be any set in A^* and let (E_n) be a sequence in A such that $E \subseteq \bigcup E_n$. Since ν is a measure and agrees with μ on A we have

$$\nu(E) \leqslant \nu\left(\bigcup_{n=1}^{\infty} E_n \right) \leqslant \sum_{n=1}^{\infty} \nu(E_n) = \sum_{n=1}^{\infty} \mu(E_n).$$

Therefore $\nu(E) \leqslant \mu^*(E)$ for any $E \in A^*$. Since μ^* and ν are additive, $\mu^*(E) + \mu^*(X \setminus E) = \nu(E) + \nu(X \setminus E)$. Since the terms on the right-hand side are finite and not greater than the corresponding terms on the left side, we infer that $\mu^*(E) = \nu(E)$ for all $E \in A^*$. This establishes the uniqueness when μ is a finite measure.

Suppose that μ is σ-finite and let (F_n) be an increasing sequence of sets in A with $\mu(F_n) < +\infty$ and $X = \bigcup F_n$. From the preceding paragraph, $\mu^*(E \cap F_n) = \nu(E \cap F_n)$ for each E in A^*.   Therefore

$$\mu^*(E) = \lim \mu^*(E \cap F_n)$$
$$= \lim \nu(E \cap F_n) = \nu(E),$$

so that μ^* and ν agree on A^*.   Q.E.D.

LEBESGUE MEASURE

We now return to the considerations that prompted the foregoing extension procedure, namely, to the generation of a measure on the real line R.   In Lemma 9.3 we saw that the set F of all finite unions of sets of the form

$$(a, b], \quad (-\infty, b], \quad (a, +\infty), \quad (-\infty, +\infty),$$

was an algebra of subsets of R and that the length function l gives a measure on this algebra F.   If we apply the extension procedure to l and F, we generate a measure space (R, F^*, l^*).   The σ-algebra F^* obtained in this construction is called the collection of **Lebesgue measurable sets** and the measure l^* on F^* is called **Lebesgue measure.**†

Although we sometimes wish to work with (R, F^*, l^*), it is often more convenient to deal with the smallest σ-algebra containing F than with all of F^*.   It is readily seen that this smallest σ-algebra is exactly the collection of Borel sets.   The restriction of Lebesgue measure to the Borel sets is called either **Borel** or **Lebesgue measure**.   Lest the reader feel that restricting to B weakens the theory by substantially lessening the collection of measurable sets and functions, we call attention to Exercise 9.K where it is seen that every Lebesgue measurable set is contained in a Borel measurable set with the same measure, and every Lebesgue measurable function is almost everywhere equal to a Borel measurable function.

† It might be thought that every subset of the real line is Lebesgue measurable, but this is not the case.   For the construction of sets which are not Lebesgue measurable, see pp.171 ff.

Sometimes it is more convenient to use a notion of the magnitude of an interval other than length. This can be treated as follows. Let g be a monotone increasing function on R to R so that $x \leqslant y$ implies that $g(x) \leqslant g(y)$. In addition, we shall assume that g is continuous on the right at every point, so that

$$g(c) = \lim_{h \to 0+} g(c + h).$$

Since g is monotone, it also follows that

$$\lim_{x \to -\infty} g(x), \qquad \lim_{x \to +\infty} g(x)$$

both exist, although they may be $-\infty$ or $+\infty$.

For such a function we define

$$\mu_g\big((a, b]\big) = g(b) - g(a),$$
$$\mu_g\big((-\infty, b]\big) = g(b) - \lim_{x \to -\infty} g(x),$$
$$\mu_g\big((a, +\infty)\big) = \lim_{x \to +\infty} g(x) - g(a),$$
$$\mu_g\big((-\infty, \infty)\big) = \lim_{x \to +\infty} g(x) - \lim_{x \to -\infty} g(x).$$

We further define μ_g on the algebra F of finite disjoint unions of such sets to be the corresponding sums. If the reader will check the details of the proof of Lemma 9.3, he will see that it can be easily modified to show that μ_g gives a σ-finite measure on the algebra F. Therefore, this measure has a unique extension, which we also denote by μ_g to the algebra of all Borel subsets of R. This extension is often referred to as the **Borel-Stieltjes measure** generated by g. (Of course, by applying Theorem 9.7, μ_g has an extension to a complete σ-algebra which contains the Borel sets. This extension is called the **Lebesgue-Stieltjes** measure generated by g.)

LINEAR FUNCTIONALS ON C

We shall conclude this chapter by showing that there is an intimate correspondence between Borel-Stieltjes measures on a finite closed interval $J = [a, b]$ and bounded positive linear functionals on the

Banach space $C(J)$ of all continuous functions on J to R with the norm

(9.8) $$\|f\| = \sup\{|f(x)| : x \in J\}.$$

This result, due to F. Riesz, has been considerably extended in many directions. Indeed, it is taken as the point of departure for the development of a theory of integration by many authors who prefer to regard the integral as a linear functional on spaces of continuous functions. We choose to take a very concrete approach to this theorem and offer a proof which is closely parallel to the Riemann–Stieltjes integral version presented in Reference [1], pp. 290–294.

9.9 RIESZ REPRESENTATION THEOREM. *If G is a bounded positive linear functional on* $C(J)$, *then there exists a measure* γ *defined on the Borel subsets of* R *such that*

(9.9) $$G(f) = \int_J f \, d\gamma$$

for all f in $C(J)$. *Moreover, the norm* $\|G\|$ *of G equals* $\gamma(J)$.

PROOF. If t is such that $a \leqslant t < b$ and n is a sufficiently large natural number, let $\varphi_{t,n}$ be the function in $C(J)$ which equals 1 on $[a, t]$, which equals 0 on $(t + 1/n, b]$, and which is linear on $(t, t + 1/n]$. If $n \leqslant m$ and $x \in J$, then $0 \leqslant \varphi_{t,m}(x) \leqslant \varphi_{t,n}(x) \leqslant 1$, so that the real sequence $(G(\varphi_{t,n}))$ is bounded and decreasing. If $t \in [a, b)$, we define

$$g(t) = \lim_{n \to +\infty} G(\varphi_{t,n}).$$

Further, set $g(t) = 0$ for $t < a$; if $t \geqslant b$, we set $g(t) = G(\varphi_b)$ where $\varphi_b(x) = 1$ for all $x \in J$. It is readily seen that g is a monotone increasing function on R.

We claim that g is continuous from the right. This is clear if $t < a$ or $t \geqslant b$. Suppose that $t \in [a, b)$ and $\varepsilon > 0$ and let

$$n > \sup\{2, \|G\|\varepsilon^{-1}\}$$

be so large that

$$g(t) \leqslant G(\varphi_{t,n}) \leqslant g(t) + \varepsilon.$$

If ψ_n is the function in $C(J)$ which equals 1 on $[a, t + n^{-2}]$, which equals 0 on $(t + n^{-1} - n^{-2}, b]$, and which is linear on

$$(t + n^{-2}, t + n^{-1} - n^{-2}],$$

then an exercise in analytic geometry shows that $\|\psi_n - \varphi_{t,n}\| \leqslant 1/n$. Therefore

$$G(\psi_n) \leqslant G(\varphi_{t,n}) + \left(\frac{1}{n}\right) \|G\| \leqslant g(t) + 2\varepsilon,$$

so that $g(t) \leqslant g(t + n^{-2}) \leqslant g(t) + 2\varepsilon$.

According to the Hahn Extension Theorem there exists a measure γ on the Borel subsets of R such that $\gamma\big((\alpha, \beta]\big) = g(\beta) - g(\alpha)$. In particular, this shows that $\gamma(E) = 0$, if $E \cap J = \emptyset$, that

$$\gamma\big([a, c]\big) = \gamma\big((a - 1, c]\big) = g(c),$$

and that $\|G\| = |G(\varphi_b)| = g(b) = \gamma(J)$.

It remains to show that equation (9.9) holds for f in $C(J)$. If $\varepsilon > 0$, since f is uniformly continuous on J, there is a $\delta(\varepsilon) > 0$ such that if $|x - y| < \delta(\varepsilon)$ and $x, y \in J$, then $|f(x) - f(y)| < \varepsilon$. Now let $a = t_0 < t_1 < \cdots < t_m = b$ be such that $\sup \{t_k - t_{k-1}\} < \frac{1}{2}\delta(\varepsilon)$ and choose n so large that $2/n < \inf \{t_k - t_{k-1}\}$ and that for $k = 1, \ldots, m$, then

(9.10) $$g(t_k) \leqslant G(\varphi_{t_k,n}) \leqslant g(t_k) + \varepsilon(m\|f\|)^{-1}.$$

We now consider functions defined on J by

$$f_1(x) = f(t_1) \varphi_{t_1,n}(x) + \sum_{k=2}^{m} f(t_k)\{\varphi_{t_k,n}(x) - \varphi_{t_{k-1},n}(x)\},$$

$$f_2(x) = f(t_1) \chi_{[t_0,t_1]}(x) + \sum_{k=2}^{m} f(t_k) \chi_{(t_{k-1},t_k]}(x).$$

Note that $f_1 \in C(J)$ and that f_2 is a step function on J. It is clear that $\sup \{|f_2(x) - f(x)| : x \in J\} \leqslant \varepsilon$ and as an exercise (or see [1], p. 292) the reader can show that $\|f_1 - f\| \leqslant \varepsilon$. Therefore we have

$$|G(f) - G(f_1)| \leqslant \varepsilon \|G\|.$$

In view of (9.10) we see that if $2 \leqslant k \leqslant m$, then

$$|G(\varphi_{t_k,n} - \varphi_{t_{k-1},n}) - \{g(t_k) - g(t_{k-1})\}| \leqslant \varepsilon(m\,\|f\|)^{-1}.$$

Apply G to f_1 and integrate f_2 with respect to γ. The inequality just obtained yields

$$\left|G(f_1) - \int_J f_2\, d\gamma\right| \leqslant \varepsilon.$$

But since f_2 lies within ε of f, we also have

$$\left|\int_J f_2\, d\gamma - \int_J f\, d\gamma\right| \leqslant \varepsilon\,\gamma(J).$$

Combining the inequalities, we arrive at the inequality

$$\left|G(f) - \int_J f\, d\gamma\right| \leqslant \varepsilon\,(2\|G\| + 1),$$

and since ε is arbitrary, we deduce (9.9).   Q.E.D.

If the reader will check the proof of Lemma 8.13, he will see that an arbitrary bounded linear functional G on $C(J)$ can be written as the difference $G^+ - G^-$ of two positive bounded linear functionals. Making use of this observation, one can extend the Riesz Representation Theorem given above to represent a bounded linear functional on $C(J)$ by means of integration with respect to a charge defined on the Borel subsets of J.

EXERCISES

9.A. Establish that the family F of all finite unions of sets of the form (9.1) is an algebra of sets in R.

9.B. Show that the family G of all finite unions of sets of the form

$$(a, b), \quad (-\infty, b), \quad (a, +\infty), \quad (-\infty, +\infty)$$

is *not* an algebra of sets in R. However, the σ-algebra generated by G is the family of Borel sets.

9.C. Show that if the set $(a, +\infty)$ is the union of a disjoint sequence of sets $(a_n, b_n]$, then

$$\sum_{n=1}^{\infty} l\big((a_n, b_n]\big) = +\infty.$$

9.D. Let X be the set of all rational numbers r satisfying $0 < r \leqslant 1$ and let A be the family of all finite unions of "half-open intervals" of the form $\{r \in X : a < r \leqslant b\}$, where $0 \leqslant a \leqslant b \leqslant 1$ and $a, b \in Q$. Show that A is an algebra of subsets of X. Moreover, every non-empty set in A is infinite. However, the σ-algebra generated by A consists of all subsets of X.

9.E. If E is a countable subset of R, then it has Lebesgue measure zero.

9.F. Let $I_n = (n, n+1]$, for $n = 0, \pm 1, \pm 2, \ldots$. If a subset E is contained in the union of a finite number of the $\{I_n\}$, then $l^*(E) < +\infty$. However, construct a Lebesgue measurable set E with $l^*(E) < +\infty$ such that $l^*(E \cap I_n) > 0$ for all n. Show that a subset E of R is Lebesgue measurable if and only if $E \cap I_n$ is Lebesgue measurable for each n.

9.G. If A is a Lebesgue measurable subset of R and $\varepsilon > 0$, show that there exists an open set $G_\varepsilon \supseteq A$ such that

$$l^*(A) \leqslant l^*(G_\varepsilon) \leqslant l^*(A) + \varepsilon.$$

9.H. If B is a Lebesgue measurable subset of R, if $\varepsilon > 0$, and if $B \subseteq I_n = (n, n+1]$, then there exists a compact set $K_\varepsilon \subseteq B$ such that

$$l^*(K_\varepsilon) \leqslant l^*(B) \leqslant l^*(K_\varepsilon) + \varepsilon.$$

(*Hint:* Apply the Exercise 9.G to $A = I_n \setminus B$.)

9.I. If A is an arbitrary Lebesgue measurable set in R, apply the preceding exercises to show that

$$l^*(A) = \inf\{l^*(G) : A \subseteq G, G \text{ open}\},$$
$$l^*(A) = \sup\{l^*(K) : K \subseteq A, K \text{ compact}\}.$$

9.J. Let $\lambda = l^*$ denote Lebesgue measure on R, and let A be a Lebesgue measurable set with $\lambda(A) < +\infty$. If $\varepsilon > 0$, there exists an open set which is the union of a finite number of open intervals such that

$$\|\chi_A - \chi_G\| = \lambda(A \triangle G) \leqslant \varepsilon.$$

Moreover, if $\varepsilon > 0$ there exists a continuous function f such that

$$\|\chi_A - f\|_1 = \int |\chi_A - f| \, d\lambda < \varepsilon.$$

9.K. Let A be a Lebesgue measurable subset of $\boldsymbol{R}$. Show that there exists a Borel measurable subset B of $\boldsymbol{R}$ such that $A \subseteq B$ and such that $l^*(B \setminus A) = 0$. (*Hint:* Consider the case where $l^*(A) < +\infty$ first.) Show that every Lebesgue measurable set is the union of a Borel measurable set (with the same measure) and a set of Lebesgue measure zero. In the terminology of Exercise 3.L, this asserts that the Lebesgue algebra is the **completion** of the Borel algebra. As a consequence of Exercise 3.N, we infer that every Lebesgue measurable function is almost everywhere equal to a Borel measurable function.

9.L. If g belongs to $L(\boldsymbol{R}, \boldsymbol{B}, \lambda)$ and $\varepsilon > 0$, then there exists a continuous function f such that

$$\|g - f\|_1 = \int |g - f| \, d\lambda < \varepsilon.$$

9.M. If $\boldsymbol{B}$ is the Borel algebra and λ is Lebesgue measure on $\boldsymbol{B}$, show (i) $\lambda(G) > 0$ for every open $G \neq \emptyset$, (ii) $\lambda(K) < +\infty$ for every compact set K, and (iii) $\lambda(x + E) = \lambda(E)$ for all $E \in \boldsymbol{B}$. (Here $x + E = \{x + y : y \in E\}$.)

9.N. Let X be a set, A an algebra of subsets of X, and μ a measure on A. If B is an arbitrary subset of X, let $\mu'(B)$ be defined to be

$$\mu'(B) = \inf \{\mu(A) : B \subseteq A \in A\}.$$

Show that $\mu'(E) = \mu(E)$ for all $E \in A$ and that $\mu^*(B) \leqslant \mu'(B)$. Moreover, $\mu^* = \mu'$ in case X is the countable union of sets with finite μ-measure. Is μ' countably subadditive in the sense of 9.5(e)?

9.O. Let X be an uncountable set and let A be the collection of sets E which are either finite or have finite complement. In the former case let $\mu(E) = 0$; in the latter, let $\mu(E) = +\infty$. Show that μ is a measure on A. Calculate the outer measure μ^* corresponding to Definition 9.4. Calculate the set function μ' defined in Exercise 9.N. Are they the same?

9.P. Let X be a set and let α be defined for arbitrary subsets of X to R and satisfy

$$0 \leqslant \alpha(E) \leqslant \alpha(E \cup F) \leqslant \alpha(E) + \alpha(F),$$

when E and F are subsets of X. Let S be the collection of all subsets E of X such that

$$\alpha(A) = \alpha(A \cap E) + \alpha(A \setminus E)$$

for all $A \subseteq X$. If $S \neq \emptyset$, it is an algebra and α is additive on S.

9.Q. It may happen that the collection S in Exercise 9.P is empty. For example, let $\alpha(E) = 1$ for all $E \subseteq X$.

9.R. Let X and A be as in Exercise 9.D, and let A_1 be the σ-algebra generated by A. Let μ_1 be the counting measure on A_1 and let $\mu_2 = 2\mu_1$. Show that $\mu_1 = \mu_2$ on A but not on A_1. (Hence the σ-finiteness hypothesis in Theorem 9.8 cannot be dropped.)

9.S. Let g be a monotone increasing and right continuous function on R to R. If μ_g is defined as at the end of this section, show that μ_g is a measure on the algebra F.

9.T. Consider the following functions defined for $x \in R$ by:

(a) $g_1(x) = 2x$,   (b) $g_2(x) = \text{Arc tan } x$,
(c) $g_3(x) = 0, x < 0$,   (d) $g_4(x) = 0, x < 0$,
$\qquad\quad = 1, x \geqslant 0$, $\qquad\quad = x, x \geqslant 0$.

Describe the Borel-Stieltjes measures determined by these functions. Which of these measures are absolutely continuous with respect to Borel measure? What are their Radon-Nikodým derivatives? Which of these measures are singular with respect to Borel measure? Which of these measures are finite? With respect to which of these measures is Borel measure absolutely continuous?

9.U. Let μ be a finite measure on the Borel sets B of R and let $g(x) = \mu\big((-\infty, x]\big)$ for $x \in R$. Show that g is monotone increasing and right continuous, and that

$$\mu\big((a, b]\big) = g(b) - g(a)$$

when $-\infty < a \leqslant b < +\infty$.   Show that $\mu(R) = \lim\limits_{x \to \infty} g(x)$.

9.V.   Let f be Riemann integrable on $[a, b]$ to R.   Then there exists a monotone increasing sequence (φ_n) and a monotone decreasing sequence (ψ_n) of step functions such that $\varphi_n(x) \leqslant f(x) \leqslant \psi_n(x)$ for $x \in [a, b]$ and

$$\lim \int \varphi_n \, d\lambda = \lim \int \psi_n \, d\lambda.$$

(Here λ denotes Lebesgue measure.)   Show that $f = \lim \psi_n = \lim \varphi_n$ almost everywhere, that f is Lebesgue measurable, and that

$$\int f \, d\lambda = \int_a^b f(x) \, dx.$$

CHAPTER 10

Product Measures

Let X and Y be two sets; then the **Cartesian product** $Z = X \times Y$ is the set of all ordered pairs (x, y) with $x \in X$ and $y \in Y$. We shall first show that the Cartesian product of two measurable spaces (X, X) and (Y, Y) can be made into a measurable space in a natural fashion. Next we shall show that if measures are given on each of the factor spaces, we can define a measure on the product space. Finally, we shall relate integration with respect to the product measure and iterated integration with respect to the measures in the factor spaces. The model to be kept in mind throughout this discussion is the plane, which we regard as the product $R \times R$.

10.1 DEFINITION. If (X, X) and (Y, Y) are measurable spaces, then a set of the form $A \times B$ with $A \in X$ and $B \in Y$ is called a **measurable rectangle**, or simply a **rectangle**, in $Z = X \times Y$. We shall denote the collection of all finite unions of rectangles by Z_0.

It is an exercise to show that every set in Z_0 can be expressed as a finite *disjoint* union of rectangles in Z (see Exercise 10.D).

10.2 LEMMA. *The collection Z_0 is an algebra of subsets of Z.*

PROOF. It is clear that the union of a finite number of sets in Z_0 also belongs to Z_0. Similarly, it follows from the first part of Exercise 10.E that the complement of a rectangle in Z belongs to Z_0. Apply De Morgan's laws to see that the complement of any set in Z_0 belongs to Z_0.   Q.E.D.

10.3 DEFINITION.   If (X, X) and (Y, Y) are measurable spaces, then $Z = X \times Y$ denotes the σ-algebra of subsets of $Z = X \times Y$ generated by rectangles $A \times B$ with $A \in X$ and $B \in Y$.   We shall refer to a set in Z as a **Z-measurable** set, or as a **measurable** subset of Z.

If (X, X, μ) and (Y, Y, ν) are measure spaces, it is natural to attempt to define a measure π on the subsets of $Z = X \times Y$ which is the "product" of μ and ν in the sense that

$$\pi(A \times B) = \mu(A)\,\nu(B), \quad A \in X, B \in Y.$$

(Recall the convention that $0(\pm\infty) = 0$.)   We shall now show that this can always be done.

10.4 PRODUCT MEASURE THEOREM.   *If (X, X, μ) and (Y, Y, ν) are measure spaces, then there exists a measure π defined on $Z = X \times Y$ such that*

(10.1) $\pi(A \times B) = \mu(A)\,\nu(B)$

for all $A \in X$ and $B \in Y$.   If these measure spaces are σ-finite, then there is a unique measure π with property (10.1).

PROOF.   Suppose that the rectangle $A \times B$ is the disjoint union of a sequence $(A_j \times B_j)$ of rectangles; thus

$$\chi_{A \times B}(x, y) = \chi_A(x)\,\chi_B(y) = \sum_{j=1}^{\infty} \chi_{A_j}(x)\,\chi_{B_j}(y)$$

for all $x \in X$, $y \in Y$.   Hold x fixed, integrate with respect to ν, and apply the Monotone Convergence Theorem to obtain

$$\chi_A(x)\,\nu(B) = \sum_{j=1}^{\infty} \chi_{A_j}(x)\,\nu(B_j).$$

A further application of the Monotone Convergence Theorem yields

$$\mu(A)\,\nu(B) = \sum_{j=1}^{\infty} \mu(A_j)\,\nu(B_j).$$

Now let $E \in Z_0$; without loss of generality we may assume that

$$E = \bigcup_{j=1}^{n} (A_j \times B_j),$$

where the sets $A_j \times B_j$ are mutually disjoint rectangles. If we define $\pi_0(E)$ by

$$\pi_0(E) = \sum_{j=1}^{n} \mu(A_j)\, \nu(B_j),$$

the argument in the previous paragraph implies that π_0 is well-defined and countably additive on $\mathbf{Z}_0$. By Theorem 9.7, there is an extension of π_0 to a measure π on the σ-algebra $\mathbf{Z}$ generated by $\mathbf{Z}_0$. Since π is an extension of π_0, it is clear that (10.1) holds.

If (X, X, μ) and (Y, Y, ν) are σ-finite, then π_0 is a σ-finite measure on the algebra $\mathbf{Z}_0$ and the uniqueness of a measure satisfying (10.1) follows from the uniqueness assertion of the Hahn Extension Theorem 9.8.   Q.E.D.

Theorem 10.4 establishes the existence of a measure π on the σ-algebra $\mathbf{Z}$ generated by the rectangles $\{A \times B : A \in X, B \in Y\}$ and such that (10.1) holds. Any such measure will be called a **product** of μ and ν. If μ and ν are both σ-finite, then they have a unique product. In the general case the extension procedure discussed in the previous chapter leads to a uniquely determined product measure. However, it will be seen in Exercise 10.S that it is possible for two distinct measures on $\mathbf{Z}$ to satisfy (10.1) if μ and ν are not σ-finite.

In order to relate integration with respect to a product measure and iterated integration, the notion of a section is useful.

10.5 DEFINITION. If E is a subset of $Z = X \times Y$ and $x \in X$, then the x-**section of** E is the set

$$E_x = \{y \in Y : (x, y) \in E\}$$

Similarly, if $y \in Y$, then the y-**section of** E is the set

$$E^y = \{x \in X : (x, y) \in E\}.$$

If f is a function defined on Z to $\bar{\mathbf{R}}$, and $x \in X$, then the x-**section of** f is the function f_x defined on Y by

$$f_x(y) = f(x, y), \qquad y \in Y.$$

Similarly, if $y \in Y$, then the y-**section** of f is the function f^y defined on X by

$$f^y(x) = f(x, y), \qquad x \in X.$$

10.6 LEMMA.   (a) *If E is a measurable subset of Z, then every section of E is measurable.*

(b) *If f is a measurable function on Z to $\bar{R}$, then every section of f is measurable.*

PROOF.   (a) If $E = A \times B$ and $x \in X$, then the x-section of E is B if $x \in A$, and is $\emptyset$ if $x \notin A$.   Therefore, the rectangles are contained in the collection E of sets in Z having the property that each x-section is measurable.   Since it is easily seen that E is a σ-algebra (see Exercise 10.I), it follows that $E = Z$.

(b) Let $x \in X$ and $\alpha \in R$, then

$$\begin{aligned} \{y \in Y : f_x(y) > \alpha\} &= \{y \in Y : f(x, y) > \alpha\} \\ &= \{(x, y) \in X \times Y : f(x, y) > \alpha\}_x. \end{aligned}$$

If f is Z-measurable, then f_x is Y-measurable.   Similarly, f^y is X-measurable.   Q.E.D.

We interpolate an important result, which is often useful in measure and probability theory, and which will be used below.   We recall (see Exercise 2.V) that a monotone class is a nonempty collection M of sets which contains the union of each increasing sequence in M and the intersection of each decreasing sequence in M.   It is easy (see Exercise 2.W) to show that if A is a nonempty collection of subsets of a set S, then the σ-algebra S generated by A contains the monotone class M generated by A.   We now show that if A is an algebra, then $S = M$.

10.7 MONOTONE CLASS LEMMA.   *If A is an algebra of sets, then the σ-algebra S generated by A coincides with the monotone class M generated by A.*

PROOF.   We have remarked that $M \subseteq S$.   To obtain the opposite inclusion it suffices to prove that M is an algebra.

If $E \in M$, define $M(E)$ to be the collection of $F \in M$ such that $E \setminus F$, $E \cap F$, $F \setminus E$ all belong to M.   Evidently $\emptyset$, $E \in M(E)$ and it is readily seen that $M(E)$ is a monotone class. Moreover, $F \in M(E)$ if and only if $E \in M(F)$.

If E belongs to the algebra A, then it is clear that $A \subseteq M(E)$. But since M is the smallest monotone class containing A, we must have $M(E) = M$ for E in A. Therefore, if $E \in A$ and $F \in M$, then $F \in M(E)$. We infer that if $E \in A$ and $F \in M$, then $E \in M(F)$ so that $A \subseteq M(F)$ for any $F \in M$. Using the minimality of M once more we conclude that $M(F) = M$ for any $F \in M$. Thus M is closed under intersections and relative complements. But since $X \in M$ it is plain that M is an algebra; since it is a monotone class, it is indeed a σ-algebra.   Q.E.D.

It follows from the Monotone Class Lemma that if a monotone class contains an algebra A, then it contains the σ-algebra generated by A.

10.8 LEMMA.   *Let* (X, X, μ) *and* (Y, Y, ν) *be* σ-*finite measure spaces. If* $E \in Z = X \times Y$, *then the functions defined by*

(10.2) $$f(x) = \nu(E_x), \quad g(y) = \mu(E^y)$$

are measurable, and

(10.3) $$\int_X f \, d\mu = \pi(E) = \int_Y g \, d\nu.$$

PROOF.   First we shall suppose that the measure spaces are finite and let M be the collection of all $E \in Z$ for which the above assertion is true. We shall show that $M = Z$ by demonstrating that M is a monotone class containing the algebra Z_0. In fact, if $E = A \times B$ with $A \in X$ and $B \in Y$, then

$$f(x) = \chi_A(x) \, \nu(B), \qquad g(y) = \chi_B(y) \, \mu(A),$$
$$\int_X f \, d\mu = \mu(A) \, \nu(B) = \int_Y g \, d\nu.$$

Since an arbitrary element of Z_0 can be written as a finite disjoint union of rectangles, it follows that $Z_0 \subseteq M$.

We now show that M is a monotone class. Indeed, let (E_n) be a monotone increasing sequence in M with union E. Therefore

$$f_n(x) = \nu\big((E_n)_x\big), \qquad g_n(y) = \mu\big((E_n)^y\big)$$

are measurable and

$$\int_X f_n \, d\mu = \pi(E_n) = \int_Y g_n \, d\nu.$$

It is clear that the monotone increasing sequences (f_n) and (g_n) converge to the functions f and g defined by

$$f(x) = \nu(E_x), \qquad g(y) = \mu(E^y).$$

If we apply the fact that π is a measure and the Monotone Convergence Theorem, we obtain

$$\int_X f \, d\mu = \pi(E) = \int_Y g \, d\nu,$$

so that $E \in M$. Since π is finite measure, it can be proved in the same way that if (F_n) is a monotone decreasing sequence in M, then $F = \bigcap F_n$ belongs to M. Therefore M is a monotone class, and it follows from the Monotone Class Lemma that $M = Z$.

If the measure spaces are σ-finite, let Z be the increasing union of a sequence of rectangles (Z_n) with $\pi(Z_n) < +\infty$ and apply the previous argument and the Monotone Convergence Theorem to the sequence $(E \cap Z_n)$.   Q.E.D.

10.9 TONELLI'S THEOREM. *Let (X, X, μ) and (Y, Y, ν) be σ-finite measure spaces and let F be a nonnegative measurable function on $Z = X \times Y$ to $\bar{R}$. Then the functions defined on X and Y by*

$$(10.4) \qquad f(x) = \int_Y F_x \, d\nu, \qquad g(y) = \int_X F^y \, d\mu,$$

are measurable and

$$(10.5) \qquad \int_X f \, d\mu = \int_Z F \, d\pi = \int_Y g \, d\nu.$$

In other symbols,

$$(10.6) \qquad \int_X \left(\int_Y F \, d\nu \right) d\mu = \int_Z F \, d\pi = \int_Y \left(\int_X F \, d\mu \right) d\nu.$$

PROOF. If F is the characteristic function of a set in Z, the assertion follows from the Lemma 10.8. By linearity, the present theorem holds for a measurable simple function. If F is an arbitrary nonnegative measurable function on Z to $\overline{R}$, Lemma 2.11 implies that there is a sequence (Φ_n) of nonnegative measurable simple functions which converges in a monotone increasing fashion on Z to F. If φ_n and ψ_n are defined by

$$(10.7) \qquad \varphi_n(x) = \int_Y (\Phi_n)_x \, dv, \qquad \psi_n(y) = \int_X (\Phi_n)^y \, d\mu,$$

then φ_n and ψ_n are measurable and monotone in n. By the Monotone Convergence Theorem, (φ_n) converges on X to f and (ψ_n) converges on Y to g. Another application of the Monotone Convergence Theorem implies that

$$\int_X f \, d\mu = \lim \int_X \varphi_n \, d\mu = \lim \int_Z \Phi_n \, d\pi$$
$$= \lim \int_Y \psi_n \, dv = \int_Y g \, dv.$$

The same theorem also shows that

$$\int_Z F \, d\pi = \lim \int_Z \Phi_n \, d\pi,$$

from which (10.5) follows. Q.E.D.

It will be seen in the exercises that Tonelli's Theorem may fail if we drop the hypothesis that F is nonnegative, or if we drop the hypothesis that the measures μ, ν are σ-finite.

Tonelli's Theorem deals with a nonnegative function on Z and affirms the equality of the integral over Z and the two iterated integrals whether these integrals are finite or equal $+\infty$. The final result considers the case where the function is allowed to take both positive and negative values, but is assumed to be integrable.

10.10 FUBINI'S THEOREM. *Let (X, X, μ) and (Y, Y, ν) be σ-finite spaces and let the measure π on $Z = X \times Y$ be the product of μ and ν. If the function F on $Z = X \times Y$ to R is integrable with respect to π, then the extended real-valued functions defined almost everywhere by*

$$(10.8) \qquad f(x) = \int_Y F_x \, dv, \qquad g(y) = \int_X F^y \, d\mu$$

have finite integrals and

(10.9)
$$\int_X f \, d\mu = \int_Z F \, d\pi = \int_Y g \, d\nu.$$

In other symbols,

(10.10)
$$\int_X \left[\int_Y F \, d\nu \right] d\mu = \int_Z F \, d\pi = \int_Y \left[\int_X F \, d\mu \right] d\nu.$$

PROOF. Since F is integrable with respect to π, its positive and negative parts F^+ and F^- are integrable. Apply Tonelli's Theorem to F^+ and F^- to deduce that the corresponding f^+ and f^- have finite integrals with respect to μ. Hence f^+ and f^- are finite-valued μ-almost everywhere, so their difference f is defined μ-almost everywhere and the first part of (10.9) is clear. The second part is similar.

Q.E.D.

Since we have chosen in Chapter 5 to restrict the use of the word "integrable" to real-valued functions, we cannot conclude that the functions f, g defined in (10.8) are integrable. However, they are almost everywhere equal to integrable functions.

It will be seen in an exercise that Fubini's Theorem may fail if the hypothesis that F is integrable is dropped.

EXERCISES

10.A. Let $A \subseteq X$ and $B \subseteq Y$. If A or B is empty, then $A \times B = \emptyset$. Conversely, if $A \times B = \emptyset$, then either $A = \emptyset$ or $B = \emptyset$.

10.B. Let $A_j \subseteq X$ and $B_j \subseteq Y$, $j = 1, 2$. If $A_1 \times B_1 = A_2 \times B_2 \neq \emptyset$, then $A_1 = A_2$ and $B_1 = B_2$.

10.C. Let $A_j \subseteq X$ and $B_j \subseteq Y$, $j = 1, 2$. Then
$$(A_1 \times B_1) \cup (A_2 \times B_2) = [(A_1 \setminus A_2) \times B_1]$$
$$\cup [(A_1 \cap A_2) \times (B_1 \cup B_2)] \cup [(A_2 \setminus A_1) \times B_2],$$

and the sets on the right side are mutually disjoint.

10.D. Let (X, X) and (Y, Y) be measurable spaces. If $A_j \in X$ and $B_j \in Y$ for $j = 1, \ldots, m$, then the set

$$\bigcup_{j=1}^{n} (A_j \times B_j)$$

can be written as the *disjoint* union of a finite number of rectangles in Z.

10.E. Let $A_j \subseteq X$ and $B_j \subseteq Y, j = 1, 2$. Then

$$(A_1 \times B_1) \setminus (A_2 \times B_2) = [(A_1 \cap A_2) \times (B_1 \setminus B_2)]$$
$$\cup [(A_1 \setminus A_2) \times B_1]$$
$$(A_1 \times B_1) \cap (A_2 \times B_2) = (A_1 \cap A_2) \times (B_1 \cap B_2).$$

10.F. If (R, B) denotes the measurable space consisting of real numbers together with the Borel sets, show that every open subset of $R \times R$ belongs to $B \times B$. In fact, this σ-algebra is the σ-algebra generated by the open subsets of $R \times R$. (In other words, $B \times B$ is the Borel algebra of $R \times R$.)

10.G. Let f and g be real-valued functions on X and Y, respectively; suppose that f is X-measurable and that g is Y-measurable. If h is defined for (x, y) in $X \times Y$ by $h(x, y) = f(x) g(y)$, show that h is $X \times Y$-measurable.

10.H. If E is a subset of R, let $\gamma(E) = \{(x, y) \in R \times R : x - y \in E\}$. If $E \in B$, show that $\gamma(E) \in B \times B$. Use this to prove that if f is a Borel measurable function on R to R, then the function F defined by $F(x, y) = f(x - y)$ is measurable with respect to $B \times B$.

10.I. Let E and F be subsets of $Z = X \times Y$, and let $x \in X$. Show that $(E \setminus F)_x = E_x \setminus F_x$. If (E_α) are subsets of Z, then

$$\left(\bigcup E_\alpha \right)_x = \bigcup (E_\alpha)_x.$$

10.J. Let (X, X, μ) be the measure space on the natural numbers $X = N$ with the counting measure defined on all subsets of $X = N$. Let (Y, Y, ν) be an arbitrary measure space. Show that a set E in $Z = X \times Y$ belongs to $Z = X \times Y$ if and only if each section E_n of E belongs to Y. In this case there is a unique product measure π, and

$$\pi(E) = \sum_{n=1}^{\infty} \nu(E_n), \qquad E \in Z.$$

A function f on $Z = X \times Y$ to R is measurable if and only if each section f_n is Y-measurable. Moreover, f is integrable with respect to π if and only if the series

$$\sum_{n=1}^{\infty} \int_Y |f_n| \, dv$$

is convergent, in which case

$$\int_Z f \, d\pi = \sum_{n=1}^{\infty} \left[\int_Y f_n \, dv \right] = \int_Y \left[\sum_{n=1}^{\infty} f_n \right] dv.$$

10.K. Let X and Y be the unit interval $[0, 1]$ and let X and Y be the Borel subsets of $[0, 1]$. Let μ be Lebesgue measure on X and let v be the counting measure on Y. If $D = \{(x, y) : x = y\}$, show that D is a measurable subset of $Z = X \times Y$, but that

$$\int v(D_x) \, d\mu(x) \neq \int \mu(D^y) \, dv(y).$$

Hence Lemma 10.8 may fail unless both of the factors are required to be σ-finite.

10.L. If F is the characteristic function of the set D in the Exercise 10.K, show that Tonelli's Theorem may fail unless both of the factors are required to be σ-finite.

10.M. Show that the example considered in Exercise 10.J demonstrates that Tonelli's Theorem holds for arbitrary (Y, Y, v) when (X, X, μ) is the set N of natural numbers with the counting measure on arbitrary subsets of N.

10.N. If $a_{mn} \geqslant 0$ for $m, n \in N$, then

$$\sum_{m=1}^{\infty} \sum_{n=1}^{\infty} a_{mn} = \sum_{n=1}^{\infty} \sum_{m=1}^{\infty} a_{mn} \quad (\leqslant +\infty).$$

10.O. Let a_{mn} be defined for $m, n \in N$ by requiring that $a_{nn} = +1$, $a_{n,n+1} = -1$, and $a_{mn} = 0$ if $m \neq n$ or $m \neq n + 1$. Show that

$$\sum_{m=1}^{\infty} \sum_{n=1}^{\infty} a_{mn} = 0, \qquad \sum_{n=1}^{\infty} \sum_{m=1}^{\infty} a_{mn} = 1,$$

so the hypothesis of integrability in Fubini's Theorem cannot be dropped.

10.P. Let f be integrable on (X, X, μ), let g be integrable on (Y, Y, ν), and define h on Z by $h(x, y) = f(x) g(y)$. If π is a product of μ and ν, show that h is π-integrable and

$$\int_Z h \, d\pi = \left[\int_X f \, d\mu \right] \left[\int_Y g \, d\nu \right].$$

10.Q. Suppose that (X, X, μ) and (Y, Y, ν) are σ-finite, and let E, F belong to $X \times Y$. If $\nu(E_x) = \nu(F_x)$ for all $x \in X$, then $\pi(E) = \pi(F)$.

10.R. Let f and g be Lebesgue integrable functions on (R, B) to R. From Exercise 10.H it follows that the function mapping (x, y) into $f(x - y) g(y)$ is measurable with respect to $B \times B$. If λ denotes Lebesgue measure on B, use Tonelli's Theorem and the fact that

$$\int_R |f(x - y)| \, d\lambda(x) = \int_R |f(x)| \, d\lambda(x)$$

to show that the function h defined for $x \in R$ by

$$h(x) = \int_R f(x - y) g(y) \, d\lambda(y)$$

is finite almost everywhere. Moreover,

$$\int |h| \, d\lambda \leqslant \left[\int |f| \, d\lambda \right] \left[\int |g| \, d\lambda \right].$$

The function h defined above is called the **convolution** of f and g and is usually denoted by $f * g$.

10.S. Let $X = R$, X be the σ-algebra of all subsets of R and let μ be defined by $\mu(A) = 0$ if A is countable, and $\mu(A) = +\infty$ if A is uncountable. We shall construct distinct products of μ with itself.

(a) If $E \in Z = X \times X$, define $\pi(E) = 0$ in case E can be written as the union $E = G \cup H$ of two sets in Z such that the x-projection of G is countable and the y-projection of H is countable. Otherwise, define $\pi(E) = +\infty$. It is evident that π is a measure on Z. If $\pi(E) = 0$, then E is contained in the union of a countable set of lines in the plane. If $A, B \in X$, show that $\pi(A \times B) = \mu(A) \mu(B)$. Hence π is a product of μ with itself.

(b) If $E \in Z$, define $\rho(E) = 0$ in case E can be written as the union $E = G \cup H \cup K$ of three sets in Z such that the x-projection of G is countable, the y-projection of H is countable, and the projection of K on the line with equation $y = x$ is countable. Otherwise, define $\rho(E) = +\infty$. Now ρ is a measure on Z, and if $\rho(E) = 0$, then E is contained in the union of a countable set of lines. Show that $\rho(A \times B) = \mu(A)\,\mu(B)$ for all $A, B \in X$; hence ρ is a product of μ with itself.

(c) Let $E = \{(x, y) : x + y = 0\}$; show that $E \in Z$. However, $\rho(E) = 0$, whereas $\pi(E) = +\infty$.

The Elements of Lebesgue Measure

CHAPTER 11

Volumes of Cells and Intervals

A **cell** in R with endpoints a, b (where $a \leq b$) is a set having one of the following four forms:

(i) $\qquad\qquad (a, b) := \{x \in R : a < x < b\},$

(ii) $\qquad\qquad [a, b] := \{x \in R : a \leq x \leq b\},$

(iii) $\qquad\qquad (a, b] := \{x \in R : a < x \leq b\},$

(iv) $\qquad\qquad [a, b) := \{x \in R : a \leq x < b\}.$

A cell with the first form is called an **open cell** with endpoints a, b. A cell having the second form is called a **closed cell** with endpoints a, b. Cells having the third or fourth forms are called **half-open** or **half-closed cells**. We note that if $a = b$, then only the closed cell is nonvoid.

An open cell in R is an open subset of R, and a closed cell is a closed subset of R; both of these types of cells are often convenient to work with because they fit into the topological scheme of R. A nonvoid half-open cell is neither open nor closed, so it does not have very much topological interest; however, half-open cells have the advantage that the intersection, the union, or the difference of two half-open cells (of the same kind) are either half-open cells, or the union of two disjoint half-open cells. Moreover, the complement of

127

a half-open cell $(a, b]$ is the union of the sets $(-\infty, a]$ and $(b, +\infty)$. It can also be written as the union of a countable collection of pairwise disjoint half-open cells of the form $(\alpha_i, \beta_i]$, where $i \in N$ and $\alpha_i, \beta_i \in R$.

Cells are also often called **intervals**; however, the sets

(v) $\qquad (-\infty, a) := \{x \in R : x < a\},$

(vi) $\qquad (-\infty, a] := \{x \in R : x \le a\},$

(vii) $\qquad [b, +\infty) := \{x \in R : b \le x\},$

(viii) $\qquad (b, +\infty) := \{x \in R : b < x\},$

(ix) $\qquad (-\infty, +\infty) := R = \{x \in R : -\infty < x < +\infty\},$

are also called **intervals** but they are *not* called cells. The intervals (v) and (vi) have the right endpoint a, and the intervals (vii) and (viii) have the left endpoint b. The intervals (v) and (viii) are open subsets of R; the intervals (vi) and (vii) are closed subsets of R. The final interval (ix) is *both* open and closed in R.

The **length** of a cell in R with endpoints $a \le b$ is defined to be equal to $b - a$. Thus all four of the cells

$$(a, b), \qquad (a, b], \qquad [a, b), \qquad (a, b]$$

have the same length. We will frequently write

$$l([a, b)) := b - a.$$

If I is any cell with endpoints $a \le b$, then the **interior** I° of I is the open cell (a, b), and the **closure** I^- of I is the closed cell $[a, b]$. Thus if I is any cell with endpoints $a \le b$, then the length of I is equal to the lengths of its interior I° and of its closure I^-; in symbols

$$l(I) = l(I^\circ) = l(I^-).$$

The length of any one of the intervals (v) – (ix) is taken to be $+\infty$; we write, for example,

$$l((-\infty, a]) = +\infty.$$

CELLS, CUBES, AND INTERVALS

If $p \in R^p$, $p > 1$, then a **cell** I in R^p is the Cartesian product of p cells $I_1, \ldots, I_p$ in R; thus

$$I = I_1 \times \ldots \times I_p.$$

An **open cell** in R^p is the Cartesian product of p open cells in R, and a **closed cell** in R^p is the Cartesian product of p closed cells in R. An open cell in R^p is an open set in R^p, and a closed cell in R^p is a closed subset of R^p. We say that a cell I in R^p is **half-open** if it is the Cartesian product of cells $I_1, \ldots, I_p$ in R all of which have the ("southwest") form $[a, b)$, or all of which have the ("northeast") form $(a, b]$. It is an exercise to show that if I and J are half-open cells in R^p having the same form, then $I \cap J$, $I \cup J$, and $I - J$ are the unions of a finite number of pairwise disjoint half-open cells in R^p. Similarly, the complement $I^c := R^p - I$ is the union of a countable collection of pairwise disjoint half-open cells in R^p.

A **cube** in R^p is a cell all of whose sides have equal length. Cubes may be open, closed, or half-open. An **interval** in R^p is the Cartesian product of p intervals in R. Thus a cell in R^p is also an interval in R^p, but not conversely.

If $I = I_1 \times \ldots \times I_p$ is a cell in R^p, then the p-**dimensional volume** $l(I)$ of I is defined to be the product of the lengths of the sides of $I_1, \ldots, I_p$. Thus, if the endpoints of I_j are $a_j \le b_j$ for $j = 1, \ldots, p$, then the volume of I is given by

$$l(I) := (b_1 - a_1) \ldots (b_p - a_p),$$

As in the case of cells in R, if $I = I_1 \times \ldots \times I_p$ is a cell in R^p and the endpoints of I_j are $a_j \le b_j$, then the **interior** I° of I in R^p is the open cell

$$I^\circ := (a_1, b_1) \times \ldots \times (a_p, b_p)$$

and the **closure** I^- of I in R^p is the closed cell

$$I^- := [a_1, b_1] \times \ldots \times [a_p, b_p].$$

Hence the volume $l(I)$ of a cell I in $\boldsymbol{R}^p$ is equal to the volume $l(I^\circ)$ of its interior, and also to the volume $l(I^-)$ of its closure.

We also note that the empty set $\emptyset$ can be considered to be an open cell with equal endpoints. Consequently, the above definition of the volume yields $l(\emptyset) = 0$.

Let $I = I_1 \times \ldots \times I_p$ be an interval in $\boldsymbol{R}^p$; then we define the **p-dimensional volume** $l(I)$ of I to be the product

$$l(I) := l(I_1) \ldots l(I_p)$$

of the lengths of the p sides. This product is to be interpreted as equaling 0 if *at least one* of the side lengths $l(I_1), \ldots, l(I_p)$ is equal to 0 (even though some of the lengths may equal $+\infty$). In addition, this product equals $+\infty$ if all of the side lengths $l(I_1), \ldots, l(I_p)$ are different from 0 and *at least one* of them equals $+\infty$.

TRANSLATION INVARIANCE

One thing worth noticing about the volume of cells in $\boldsymbol{R}^p$ is that it is independent of the "location" of the cells in the space, and depends only on the lengths of the sides. We express this idea by saying that the volume function is "translation invariant". To clarify this notion, we introduce the following definition.

11.1   DEFINITION.   If A is a subset of $\boldsymbol{R}^p$ and x is any vector in $\boldsymbol{R}^p$, then the **translation of A by x** is defined to be the set

$$x \oplus A := \{x + a : a \in A\}.$$

It is easy to see that the translation of a cell I in $\boldsymbol{R}^p$ is also a cell; moreover, it is an exercise to show that $l(x \oplus I) = l(I)$ for all vectors x and cells I in the space $\boldsymbol{R}^p$.

CHAPTER 12

The Outer Measure

We wish to extend the notion of the volume of a set in R^p to sets that are more general than cells or intervals. Of course, if a set E in R^p can be obtained as a disjoint union of a finite number of cells, it is natural to define the volume of E to be the sum of the volumes of the corresponding cells. However, such a set E may be decomposed in *many* ways as the disjoint union of cells and it is not immediately clear that different decompositions of E would always lead to the same value for the volume of E. In addition, many — in fact, most — sets cannot be obtained as a *finite* union of cells; the circular disk $D := \{(x, y) \in R^2 : x^2 + y^2 \leq 1\}$ is an important example of such a set. It is possible to show that D can be obtained as the disjoint union of a countable family of cells in R^2; in fact, this can be done in *infinitely many* ways. To define the 2-dimensional volume (that is, the "area") of D, we would surely need to use an infinite series for each decomposition, and to show that different decompositions of D give the same 2-dimensional volume for D. Even for as simple a figure as the disk D, this would not be an easy task.

Our approach in generalizing the notion of volume to an *arbitrary* subset of R^p will be by a process of "overestimation". Basically, we

131

will enclose the set in the union of a countable collection of cells, and then minimize the sum of the volumes of the cells by taking the infimum over all such countable collections. (This process is rather similar to the construction of the upper integral that is often done in elementary calculus courses.) This "generalized volume" is useful for many purposes and has the advantage that it is defined for *all* subsets of R^p. Unfortunately, it will be shown later that this generalized volume is not sufficiently well-behaved on the collection of *all* subsets of R^p. However, when this generalized volume is restricted to a suitable collection (to be denoted by $\mathcal{L}$) of subsets of R^p, it will be seen to have some very satisfactory properties. In Chapter 4, this collection $\mathcal{L}$ will be shown to contain all of the subsets of R^p that arise in ordinary work. In particular, $\mathcal{L}$ contains all open subsets and all closed subsets of R^p, and it contains all countable unions, all countable intersections, and all complements of sets that belong to the collection $\mathcal{L}$.

12.1 DEFINITION.   If $E \subseteq R^p$, we define the **outer measure** $m^*(E)$ of E to be

(12.1)
$$m^*(E) := \inf \left\{ \sum_{k=1}^{+\infty} l(I_k) \right\},$$

where the infimum is extended over all sequences (I_k) of cells in R^p that cover E in the sense that

(12.2)
$$E \subseteq \bigcup_{k=1}^{+\infty} I_k.$$

(1) We note that, since the entire set R^p is contained in the union of a sequence (I_k) of cells in R^p, the collection of all sequences that satisfy (12.2) is not empty. Therefore, the infimum appearing in (12.1) is well-defined and it is evident that $m^*(E) \geq 0$. We also note that it is certainly possible to have $m^*(E) = +\infty$; for example, this is the case when we take $E = R^p$.

(2) The terms $l(I_k)$ certainly satisfy $0 \le l(I_k) < +\infty$ for all k. Hence, the series

(12.3)
$$\sum_{k=1}^{+\infty} l(I_k)$$

is either (i) absolutely convergent, in which case the value of its sum does not depend on the order of summation, or (ii) the series (12.3) is divergent ($=$ convergent to $+\infty$), in which case we assign the value $+\infty$ as its sum.

(3) Since the convergence or the divergence of (12.3) does not depend on the order of the summands, it is equally appropriate to think of (I_k) as being a *countable family* of cells rather than a *sequence* of cells.

Before we proceed, we wish to make some further observations about Definition 12.1 that will be important.

12.2 REMARKS. (a) We may restrict the cells (I_k) in Definition 12.1 to be *closed* cells if we wish to do so. Indeed, since $I_k \subseteq I_k^-$, it is trivial that if (12.2) holds, then we also have $E \subseteq \bigcup_{k=1}^{+\infty} I_k^-$. Since, as was noted in the preceding chapter, we have $l(I_k^-) = l(I_k)$, the value of $m^*(E)$ is not affected by using the closures of the cells.

(b) We may also restrict the cells (I_k) to be *open* cells if we wish to do so. Indeed, suppose that $m^*(E) < +\infty$ and that (12.2) holds for a sequence (I_k) of cells. If $\varepsilon > 0$, then let (J_k) be a sequence of open cells such that

$$I_k \subseteq J_k \qquad \text{and} \qquad l(J_k) \le l(I_k) + \varepsilon/2^k$$

for $k = 1, 2, \ldots$. Then we have

$$E \subseteq \bigcup_{k=1}^{+\infty} J_k \qquad \text{and} \qquad \sum_{k=1}^{+\infty} l(J_k) \le \sum_{k=1}^{+\infty} l(I_k) + \varepsilon.$$

Hence it follows (why?) that the same result is obtained in (12.1) by using open cells as using arbitrary cells.

(c) If desired, we may restrict the cells (I_k) to be *half-open* cells, having either form.

(d) If $\delta > 0$ is given, we may restrict the cells (I_k) to have diameter less than δ. [Recall that if $A \subseteq \mathbf{R}^p$, then the **diameter** of A is defined to be

$$\mathrm{diam}(A) := \sup\{\|x - y\| : x, y \in A\}.$$

Hence, if $x, y \in A$, then we have $\|x-y\| \leq \mathrm{diam}(A)$.] To see that this is true, note that any half-open cell can be obtained as the union of a finite collection of pairwise disjoint half-open cells with diameters less than δ. For example, one can successively bisect the sides of the cell to obtain cells with diameter less than the preassigned δ.

We shall now establish the basic properties of the outer measure function m^*.

12.3 THEOREM.   *The outer measure function m^* defined in Definition 12.1 satisfies:*

(i) $0 \leq m^*(E) \leq +\infty$ for all $E \subseteq \mathbf{R}^p$, and $m^*(\emptyset) = 0$;
(ii) *if* $E \subseteq F$, *then* $m^*(E) \leq m^*(F)$;
(iii) *if* (E_k) *is a sequence of subsets of* $\mathbf{R}^p$, *then*

$$m^*\left(\bigcup_{k=1}^{+\infty} E_k\right) \leq \sum_{k=1}^{+\infty} m^*(E_k).$$

PROOF. (i) The first property has already been noted. If we take $I_k = \emptyset$ for all $k \in \mathbf{N}$, then $\emptyset \subseteq I_1 \cup I_2 \cup \ldots$, so

$$0 \leq m^*(\emptyset) \leq 0 + 0 + \ldots = 0.$$

Therefore (i) is proved.

(ii) If (I_k) is a sequence of cells with $F \subseteq \bigcup_{k=1}^{+\infty} I_k$, then also $E \subseteq \bigcup_{k=1}^{+\infty} I_k$. Therefore $m^*(E) \leq m^*(F)$, proving (ii).

(iii) It suffices to prove the assertion in the case that $m^*(E_n) < +\infty$ for each $n \in \mathbf{N}$. (Why?) Let $\varepsilon > 0$ and, for each $n \in \mathbf{N}$, choose a sequence $(I_k^n)_k$ of cells such that

$$E_n \subseteq \bigcup_{k=1}^{+\infty} I_k^n \qquad \text{and} \qquad \sum_{k=1}^{+\infty} l(I_k^n) \leq m^*(E_n) + \varepsilon/2^n.$$

Since $\{I_k^n : k, n \in \mathbf{N}\}$ is a countable family of cells that cover the set $\bigcup_{n=1}^{+\infty} E_n$, it follows from the definition of m^* that

$$m^* \left(\bigcup_{n=1}^{+\infty} E_n \right) \leq \sum_{k,n=1}^{+\infty} l(I_k^n) = \sum_{n=1}^{+\infty} \sum_{k=1}^{+\infty} l(I_k^n)$$

$$\leq \sum_{n=1}^{+\infty} \left(m^*(E_n) + \varepsilon/2^n \right) = \sum_{n=1}^{+\infty} m^*(E_n) + \varepsilon.$$

[Since $m^*(E) \geq 0$ for any set $E \subseteq \mathbf{R}^p$, the change from a double sum to an iterated sum is justified.] Now, since $\varepsilon > 0$ is arbitrary, property (iii) is proved. Q.E.D.

Property (iii) of Theorem 12.3 is referred to by saying that m^* is **countably subadditive** on the collection of all subsets of $\mathbf{R}^p$. In particular, it follows from (iii) that if A and B are *disjoint* subsets (that is, if $A \cap B = \emptyset$), then

$$m^*(A \cup B) \leq m^*(A) + m^*(B).$$

Our intuitive feeling about volume is that *equality* should hold in this relation. Unfortunately, it is not the case (as we will prove in a later chapter) that equality holds for an arbitrary pair of disjoint subsets of $\mathbf{R}^p$. However, we now show that this desired equality relation is true in case the sets A and B are at a *positive distance* from each other.

12.4 THEOREM. *Let A and B be disjoint subsets of $\mathbf{R}^p$ with* $\mathrm{dist}(A, B) := \inf\{\|a - b\| : a \in A, b \in B\} > 0$. *Then we have*

$$m^*(A \cup B) = m^*(A) + m^*(B).$$

PROOF. We have already seen that it is always the case that $m^*(A \cup B) \leq m^*(A) + m^*(B)$. Therefore (why?), it suffices to prove the opposite inequality under the hypothesis that $m^*(A \cup B) < +\infty$ and $\delta := \mathrm{dist}(A, B) > 0$.

If $\varepsilon > 0$ is given, let (I_n) be a covering of $A \cup B$ such that

$$\sum_{n=1}^{+\infty} l(I_n) \leq m^*(A \cup B) + \varepsilon.$$

As was noted in Remark 12.2(d), we may assume that the cells (I_n) have diameter less than δ. In this case, no cell I_n can contain both points in A and points in B, for then we would have dist$(A, B) < \delta$, contrary to the hypothesis. Hence we may divide the cells $\{I_n\}$ into three classes: (i) the cells $\{J_j\}$ that contain points in A, (ii) the cells $\{K_k\}$ that contain points in B, and (iii) the cells $\{H_h\}$ that do not contain points in either A or in B. Therefore, we have

$$m^*(A) \leq \sum_j l(J_j) \qquad \text{and} \qquad m^*(B) \leq \sum_k l(K_k),$$

from which it follows that

$$m^*(A) + m^*(B) \leq \sum_j l(J_j) + \sum_k l(K_k) + \sum_h l(H_h)$$

$$\leq \sum_n l(I_n) \leq m^*(A \cup B) + \varepsilon.$$

Therefore, we have $m^*(A) + m^*(B) \leq m^*(A \cup B) + \varepsilon$. Since $\varepsilon > 0$ is arbitrary, we infer that $m^*(A) + m^*(B) \leq m^*(A \cup B)$, as was to be shown.   Q.E.D.

It is conceivable that the value m^* gives to a cell might be different from the p-dimensional volume of the cell; however, we will now show that such an unfortunate situation cannot occur.

12.5   THEOREM.   *If I is any cell in $\mathbf{R}^p$, then $m^*(I) = l(I)$.*

PROOF.   Since the sequence $(I, \emptyset, \emptyset, \ldots)$ is a covering of I, it follows that $m^*(I) \leq l(I) + 0 + 0 + \ldots = l(I)$. To establish the opposite inequality, let $\varepsilon > 0$ be given and let (I_k) be a covering of I by open cells such that

$$\sum_{k=1}^{+\infty} l(I_k) \leq m^*(I) + \varepsilon.$$

If I is a closed cell, let $J := I$; otherwise, let J be a closed cell such that $J \subseteq I$ and $l(I) - \varepsilon < l(J)$. By the Heine–Borel Theorem there is an $m \in \boldsymbol{N}$ such that $J \subseteq \bigcup_{k=1}^{m} I_k$.

We now divide the space $\boldsymbol{R}^p$ into a finite number of closed intervals by extending the $(p - 1)$-dimensional hyperplanes that contain a face of one of the cells $I_1, \ldots, I_m$, and of J. Let $K_1, \ldots, K_n$ be the distinct closed cells into which the cells $I_1^-, \ldots, I_m^-$ are divided by these hyperplanes; further, let $J_1, \ldots, J_r$ be the closed cells into which J is divided. Therefore, we have

$$l(J) = \sum_{j=1}^{r} l(J_j) \le \sum_{k=1}^{n} l(K_k)$$

$$\le \sum_{k=1}^{m} l(I_k) \le m^*(I) + \varepsilon.$$

Consequently, $l(I) \le l(J) + \varepsilon \le m^*(I) + 2\varepsilon$. Since $\varepsilon > 0$ is arbitrary, we deduce that $l(I) \le m^*(I)$. Therefore it follows that $l(I) = m^*(I)$, as claimed.   Q.E.D.

TRANSLATION INVARIANCE

We noted at the end of Chapter 11 that the volume of a cell is not changed by a translation. We now show that the analogous result holds for the outer measure of an arbitrary subset of $\boldsymbol{R}^p$.

12.6 THEOREM. If $E \subseteq \boldsymbol{R}^p$ and $x \in \boldsymbol{R}^p$, then we have

$$m^*(x \oplus E) = m^*(E).$$

PROOF. If the sequence (I_k) of cells is a covering for E, then the sequence $(x \oplus I_k)$ of x-translates is a covering of $x \oplus E$ by cells. Since we have $\sum_k l(I_k) = \sum_k l(x \oplus I_k)$, it follows that

$$m^*(x \oplus E) \le \inf \left\{ \sum_{k=1}^{+\infty} l(x \oplus I_k) \right\}$$

$$= \inf \left\{ \sum_{k=1}^{+\infty} l(I_k) \right\} = m^*(E).$$

Since every covering (J_k) of the set $x \oplus E$ by cells can be obtained by translating a covering (K_k) of E (how?), it also follows that we have $m^*(E) \leq m^*(x \oplus E)$, whence $m^*(E) = m^*(x \oplus E)$. We leave the precise details as an exercise.   Q.E.D.

CHAPTER 13

Measurable Sets

In the preceding chapter we defined the outer measure $m^*(E)$ of an arbitrary subset E of R^p. While the function m^* has the distinct advantage of being defined for *every* subset of R^p, it is not always additive over disjoint subsets; that is, it does not always satisfy

$$m^*(A \cup B) = m^*(A) + m^*(B)$$

when A and B are sets such that $A \cap B = \emptyset$.

In this chapter we will prove that, by restricting m^* to a certain family $\mathcal{L}$ of subsets of R^p, we obtain a function that not only is additive over disjoint sets, but is even *countably additive* in the sense that if (E_n) is a sequence of sets in $\mathcal{L}$ that are pairwise disjoint (i.e., $E_n \cap E_m = \emptyset$ for $n \neq m$), then the union $E := \bigcup_{n=1}^{+\infty} E_n$ belongs to $\mathcal{L}$ and

$$(13.1) \qquad m^*(E) = \sum_{n=1}^{+\infty} m^*(E_n).$$

This countable additivity property is a very desirable one; in fact, it is this property that makes many of the properties of the Lebesgue theory of integration work nicely.

The main theorem in this chapter is due to the well-known Greco-German mathematician Constantin Carathéodory. We will state this result for the outer measure m^* introduced in Chapter 12, although his theorem remains true for any function satisfying the properties stated in Theorem 12.3.

In order to simplify the statement of Carathéodory's theorem, we introduce some terminology that is often useful.

13.1   DEFINITION. Let X be an arbitrary set. Then a family Σ of subsets of X is said to be a σ-**algebra in** X if the following conditions are satisfied:

(i) $\emptyset$ and X belong to Σ;

(ii) if $E \in \Sigma$, then the complement $E^c = X - E$ belongs to Σ;

(iii) if (E_n) is a sequence of sets in Σ, then the union $\bigcup_{n=1}^{+\infty} E_n$ belongs to Σ.

REMARKS. (a)   If Σ is a σ-algebra of subsets of X, then the intersection of a sequence of sets in Σ also belongs to Σ.

(b) If X is any set, then $\Sigma_1 := \{\emptyset, X\}$ is a (rather trivial) example of a σ-algebra of subsets of X.

(c) If X is any set and E is a subset of X, then $\Sigma_2 := \{\emptyset, E, E^c, X\}$ is a σ-algebra of subsets of X.

(d) If X is any set, then $\Sigma_3 := \{$all subsets of $X\}$ is a σ-algebra of subsets of X.

(e) If X is a set and Σ_1 and Σ_2 are σ-algebras of X, then $\Sigma_1 \cap \Sigma_2$ ($=$ the collection of all subsets of X that belong to both Σ_1 and Σ_2) is also a σ-algebra of subsets of X.

13.2   DEFINITION. Let X be a set and let Σ be a σ-algebra of subsets of X. Then an extended real-valued function μ defined on Σ is said to be a **measure on** Σ in case it satisfies:

(i) $\mu(\emptyset) = 0$;

(ii) $0 \leq \mu(E) \leq +\infty$ for all $E \in \Sigma$;

(iii) if (E_n) is a sequence of sets in Σ that are pairwise disjoint, then we have the equation

$$(13.2) \qquad \mu\left(\bigcup_{n=1}^{+\infty} E_n\right) = \sum_{n=1}^{+\infty} \mu(E_n).$$

REMARKS. (a) Since $\mu(E) \geq 0$, the series on the right side of equation (13.2) is either absolutely convergent, or it is properly divergent to $+\infty$.

(b) If $X = N$ and $\Sigma = \{$all subsets of $X\}$, then define $\mu(E)$ to be the number of elements in E if E is a finite set, and $\mu(E) := +\infty$ if E is an infinite set. Then μ is a measure on Σ and is called the **counting measure** on N.

We now introduce the "test for measurability" of a set $E \subseteq R^p$.

13.3   DEFINITION.   Let m^* be the outer measure defined on all subsets of R^p. A set $E \subseteq R^p$ is said to satisfy the **Carathéodory condition** in case

$$(13.3) \qquad m^*(A) = m^*(A \cap E) + m^*(A \cap E^c)$$

for all $A \subseteq R^p$. The collection of all such sets will be denoted by $\mathcal{L}$.

We note that the condition (13.3) can also be written in the form

$$m^*(A) = m^*(A \cap E) + m^*(A - E).$$

Intuitively, a set E satisfies the Carathéodory condition if E and its complement E^c split every set A "additively". The sets that satisfy the condition (13.3) are precisely the desired "measurable sets". The next lemma shows that the task of showing that a set E satisfies (13.3) can be simplified somewhat.

13.4 LEMMA.   *A set E satisfies the Carathéodory condition if and only if, for each set A with $m^*(A) < +\infty$, then*

$$(13.4) \qquad m^*(A) \geq m^*(A \cap E) + m^*(A \cap E^c).$$

PROOF. Indeed, since $A \cap E$ and $A \cap E^c$ are disjoint and have union A, it follows from Theorem 12.3(iii) that we always have the inequality

$$m^*(A) \leq m^*(A \cap E) + m^*(A \cap E^c).$$

Hence, if (13.4) is satisfied, then so is (13.3). Finally, we note that (13.4) is trivial in case $m^*(A) = +\infty$, so it is only necessary to treat the case $m^*(A) < +\infty$.   Q.E.D.

We will now obtain Carathéodory's theorem.

13.5 THEOREM. (Carathéodory) *Let m^* be the outer measure defined in Definition 12.1. Then the set $\mathcal{L}$ of all subsets of $\mathbf{R}^p$ that satisfy the Carathéodory Condition 13.3 is a σ-algebra of subsets of $\mathbf{R}^p$. Moreover, the restriction of m^* to $\mathcal{L}$ is a measure on $\mathcal{L}$.*

PROOF. It is clear that $\emptyset$ satisfies (13.3), and that if E satisfies (13.3), then so does its complement E^c. Hence the family of sets that satisfy the Carathéodory condition satisfies properties (i) and (ii) of Definition 13.1.

We now show that if E and F satisfy (13.3), then so does $E \cap F$. For, since $E \in \mathcal{L}$, then

$$m^*(A) = m^*(A \cap E) + m^*(A \cap E^c)$$

for any $A \subseteq \mathbf{R}^p$. Since $F \in \mathcal{L}$ we have

$$m^*(A \cap E) = m^*(A \cap E \cap F) + m^*(A \cap E \cap F^c),$$

whence it follows that

$$m^*(A) = m^*(A \cap E \cap F) + m^*(A \cap E \cap F^c) + m^*(A \cap E^c).$$

But since $E \in \mathcal{L}$, we also have

$$
\begin{aligned}
m^*\big(A \cap (E \cap F)^c\big) \\
= m^*\big(A \cap (E \cap F)^c \cap E\big) + m^*\big(A \cap (E \cap F)^c \cap E^c\big) \\
= m^*(A \cap F^c \cap E) + m^*(A \cap E^c).
\end{aligned}
$$

It therefore follows that

$$m^*(A) = m^*\big(A \cap (E \cap F)\big) + m^*\big(A \cap (E \cap F)^c\big)$$

for all sets A. Consequently, $E \cap F$ belongs to $\mathcal{L}$.

Since $\mathcal{L}$ contains the complements of sets in $\mathcal{L}$, it follows from De Morgan's Laws that if $E, F \in \mathcal{L}$, then $E \cup F \in \mathcal{L}$. Moreover, if $E \cap F = \emptyset$, then it follows from the fact that E satisfies (13.3) with A replaced by $A \cap (E \cup F)$ and $F = F \cap E^c$ that

$$\begin{aligned} m^*\big(A \cap (E \cup F)\big) &= m^*\big(A \cap (E \cup F) \cap E\big) + m^*\big(A \cap (E \cup F) \cap E^c\big) \\ &= m^*(A \cap E) + m^*(A \cap F). \end{aligned}$$

By induction we conclude that if $E_1, \ldots, E_n$ belong to $\mathcal{L}$ and are pairwise disjoint, then $E_1 \cup \ldots \cup E_n$ belongs to $\mathcal{L}$ and

$$m^*(A \cap (E_1 \cup \ldots \cup E_n)) = m^*(A \cap E_1) + \ldots + m^*(A \cap E_n)$$

for all $A \subseteq \mathbf{R}^p$.

We now show that $\mathcal{L}$ is a σ-algebra and that m^* is countably additive on $\mathcal{L}$. To do this, let (E_n) be a pairwise disjoint sequence in $\mathcal{L}$ and let $E := \bigcup_{k=1}^{+\infty} E_k$. We know that $F_n := \bigcup_{k=1}^{n} E_k$ belongs to $\mathcal{L}$ for all $n \in \mathbf{N}$. Moreover, if $A \subseteq \mathbf{R}^p$, then

$$\begin{aligned} m^*(A) &= m^*(A \cap F_n) + m^*(A \cap F_n^c) \\ &= m^*\left(\bigcup_{k=1}^{n} A \cap E_k \right) + m^*(A \cap F_n^c) \\ &= \sum_{k=1}^{n} m^*(A \cap E_k) + m^*(A \cap F_n^c). \end{aligned}$$

Since $F_n \subseteq E$, then $A \cap F_n^c \supseteq A \cap E^c$ for all $n \in \mathbf{N}$, so that

$$m^*(A) \ge \sum_{k=1}^{n} m^*(A \cap E_k) + m^*(A \cap E^c),$$

which implies that

$$(13.5) \qquad m^*(A) \geq \sum_{k=1}^{+\infty} m^*(A \cap E_k) + m^*(A \cap E^c).$$

On the other hand, it follows from the countable subadditivity of m^* that

$$(13.6) \qquad m^*(A \cap E) = m^* \left(\bigcup_{k=1}^{+\infty} A \cap E_k \right) \leq \sum_{k=1}^{+\infty} m^*(A \cap E_k).$$

Therefore, we have

$$m^*(A) \geq m^*(A \cap E) + m^*(A \cap E^c),$$

which (in view of Lemma 13.4) implies that $E \in \mathcal{L}$, so that $\mathcal{L}$ is a σ-algebra. In addition, if we take $A = E$ in (13.5) and (13.6), we obtain

$$m^*(E) = \sum_{k=1}^{+\infty} m^*(E_k),$$

which shows that m^* is countably additive on $\mathcal{L}$.   Q.E.D.

13.6   DEFINITION. If m^* is the outer measure defined in Definition 12.1, then the σ-algebra $\mathcal{L}$ of subsets of R^p that satisfy the Carathéodory Condition 13.3 is called the **Lebesgue σ-algebra** of R^p. A set $E \in \mathcal{L}$ is called a **Lebesgue measurable subset** of R^p, or briefly, a **measurable subset** of R^p. The restriction m of m^* to $\mathcal{L}$ is called **Lebesgue measure** on R^p.

Since m is the restriction of m^* to the σ-algebra $\mathcal{L}$, we have $m(E) = m^*(E)$ for all $E \in \mathcal{L}$. Ordinarily, when we know that a set E is measurable, we will write $m(E)$ instead of $m^*(E)$.

The next result assures that cells in R^p are measurable sets and that their Lebesgue measure coincides with their volume.

13.7   THEOREM. *If I is a cell in R^p, then I is measurable, and hence $m(I) = l(I)$.*

PROOF. We will give the proof for an open cell, leaving to the reader to extend the conclusion to an arbitrary cell in R^p. It was seen in Lemma 13.4 that it suffices to show that if $A \subseteq R^p$ is such that $m^*(A) < +\infty$, then

$$m^*(A) \geq m^*(A \cap I) + m^*(A - I).$$

Let $n \in N$ and let $I_n := \{x \in I : \operatorname{dist}(x, I^c) > 1/n\}$, so that $I_n \subseteq I$. Moreover, since $I - I_n$ is contained in the union of $2p$ cells each of which has one side with length $1/n$, then $m^*(I - I_n) \to 0$ as $n \to +\infty$.

Note that $A \supseteq (A \cap I_n) \cup (A - I)$ and $\operatorname{dist}(A \cap I_n, A - I) \geq 1/n$. Therefore, it follows from Theorem 12.4 that

(13.7)
$$\begin{aligned} m^*(A) &\geq m^*\big((A \cap I_n) \cup (A - I) \big) \\ &= m^*(A \cap I_n) + m^*(A - I). \end{aligned}$$

On the other hand, since

$$A \cap I = (A \cap I_n) \cup \big(A \cap (I - I_n) \big),$$

it follows from the subadditivity and the monotone character of m^* that

$$m^*(A \cap I_n) \leq m^*(A \cap I) \leq m^*(A \cap I_n) + m^*(I - I_n).$$

Therefore, we have

$$m^*(A \cap I) = \lim_{n \to \infty} m^*(A \cap I_n).$$

Hence, taking the limit in (13.7), we have

$$m^*(A) \geq m^*(A \cap I) + m^*(A - I),$$

which shows that I is a measurable set, by Lemma 13.4.   Q.E.D.

We have been able to obtain a measure defined on a σ-algebra $\mathcal{L}$ of sets that agrees with the volume function l, originally defined only for cells. Consequently, we have been successful in extending l to a larger collection of sets, and we will soon see that the collection $\mathcal{L}$ is a very large collection of sets. It is conceivable, however, that there may be *another* measure defined on $\mathcal{L}$ that agrees with l on cells. We now show that this is not the case.

   13.8   THEOREM.   *If μ is a measure defined on $\mathcal{L}$ that is such that $\mu(I) = l(I)$ for all open cells $I \subseteq \boldsymbol{R}^p$, then $\mu = m$.*

   PROOF.   For $n \in \boldsymbol{N}$, let I_n be the open cell

$$ I_n := (-n, n) \times \ldots \times (-n, n). $$

Let $E \in \mathcal{L}$ be any set with $E \subseteq I_n$ and let (J_k) be a sequence of open cells such that $E \subseteq \bigcup_{k=1}^{+\infty} J_k$. Since μ is a measure and $\mu(J_k) = l(J_k)$ for all $k \in \boldsymbol{N}$, we have

$$ \mu(E) \le \mu\left(\bigcup_{k=1}^{+\infty} J_k \right) \le \sum_{k=1}^{+\infty} \mu(J_k) = \sum_{k=1}^{+\infty} l(J_k). $$

Therefore, we have $\mu(E) \le m^*(E) = m(E)$ for all measurable sets $E \subseteq I_n$. Since μ and m are additive, then

$$ \mu(E) + \mu(I_n - E) = \mu(I_n) = m(I_n) = m(E) + m(I_n - E). $$

Since all of these terms are finite and $\mu(E) \le m(E)$ and $\mu(I_n - E) \le m(I_n - E)$, it follows that $\mu(E) = m(E)$ for all measurable sets $E \subseteq I_n$.

   Now an arbitrary measurable set E can be written as the union of a disjoint sequence (E_n) of sets, defined by

$$ E_1 := E \cap I_1, \qquad E_n := E \cap (I_n - I_{n-1}) \quad \text{for} \quad n > 1. $$

Since $\mu(E_n) = m(E_n)$ for all $n \in \boldsymbol{N}$, it follows that

$$ \mu(E) = \sum_{n=1}^{+\infty} \mu(E_n) = \sum_{n=1}^{+\infty} m(E_n) = m(E). $$

Thus μ and m agree on all measurable sets.   Q.E.D.

We conclude this chapter with two useful results that are simple properties of the positivity and the countable additivity of Lebesgue measure.

13.9 THEOREM. *If E and F are Lebesgue measurable sets and if $E \subseteq F$, then $m(E) \leq m(F)$. If, in addition, $m(E) < +\infty$, then $m(F - E) = m(F) - m(E)$.*

PROOF. Since m is additive, it is immediate from the fact that $F = E \cup (F - E)$ and $E \cap (F - E) = \emptyset$ that

$$m(F) = m(E) + m(F - E).$$

Since $m(F - E) \geq 0$, we have $m(F) \geq m(E)$. If $m(E) < +\infty$, then we can subtract $m(E)$ from both sides of the above equation. Q.E.D.

13.10 THEOREM. (a) *If (E_k) is an increasing sequence of measurable sets, then*

(13.8) $$m \left(\bigcup_{k=1}^{+\infty} E_k \right) = \lim_{k \to \infty} m(E_k).$$

(b) *If (F_k) is a decreasing sequence of Lebesgue measurable sets and if $m(F_1) < +\infty$, then*

(13.9) $$m \left(\bigcap_{k=1}^{+\infty} F_k \right) = \lim_{k \to \infty} m(F_k).$$

PROOF. (a) If $m(E_k) = +\infty$ for some $k \in \mathbf{N}$, then both sides of (13.8) are equal to $+\infty$. Hence we may suppose that $m(E_k) < +\infty$ for all $k \in \mathbf{N}$. Now let $A_1 := E_1$ and $A_k := E_k - E_{k-1}$ for $k > 1$. Then (A_k) is a disjoint sequence of measurable sets such that

$$E_k = \bigcup_{j=1}^{k} A_j \quad \text{and} \quad \bigcup_{k=1}^{+\infty} E_k = \bigcup_{k=1}^{+\infty} A_k.$$

Since m is countably additive, then

$$m\left(\bigcup_{k=1}^{+\infty} E_k\right) = m\left(\bigcup_{k=1}^{+\infty} A_k\right)$$

$$= \sum_{k=1}^{+\infty} m(A_k) = \lim_{n\to\infty}\left(\sum_{k=1}^{n} m(A_k)\right).$$

By Theorem 13.9, we have $m(A_k) = m(E_k) - m(E_{k-1})$ for $k > 1$, so the finite sum telescopes and

$$\sum_{k=1}^{n} m(A_k) = m(E_n).$$

Hence equation (13.8) is proved.

(b) Let $E_k := F_1 - F_k$ for $k \in N$, so that (E_k) is an increasing sequence of measurable sets. If we apply part (a) and Theorem 13.9, we infer that

$$m\left(\bigcup_{k=1}^{+\infty} E_k\right) = \lim_{n\to\infty} m(E_n) = \lim_{n\to\infty}\left[m(F_1) - m(F_n)\right]$$

$$= m(F_1) - \lim_{n\to\infty} m(F_n).$$

But since $\bigcup_{k=1}^{+\infty} E_k = F_1 - \bigcap_{k=1}^{+\infty} F_k$, it follows from Theorem 13.9 that

$$m\left(\bigcup_{k=1}^{+\infty} E_k\right) = m(F_1) - m\left(\bigcap_{k=1}^{+\infty} F_k\right).$$

We now combine the last two relations to obtain (13.9).   Q.E.D.

CHAPTER 14

Examples of Measurable Sets

We have established, in the preceding chapter, that there exists a σ-algebra $\mathcal{L}$ of subsets, called the **Lebesgue measurable sets**, containing all cells in $\mathbf{R}^p$, and on which there is a measure function, called **Lebesgue measure**, that extends to the sets in $\mathcal{L}$ the notion of "volume". It was asserted that the set $\mathcal{L}$ of measurable sets is a very large one, but this assertion has not yet been proved. The purpose of this chapter is to give some indications about what sets belong to this collection. We will show that all of the familiar sets in analysis are Lebesgue measurable.

14.1 LEMMA. *Every open subset of $\mathbf{R}^p$ is the union of a countable collection of open cells.*

PROOF. Let $\mathbf{Q}$ denote the set of all rational numbers in $\mathbf{R}$ and let $\mathbf{Q}^p := \mathbf{Q} \times \ldots \times \mathbf{Q}$ denote the collection of those points in $\mathbf{R}^p$ all of whose coordinates are rational numbers. It is well known that $\mathbf{Q}^p$ is a countable set; we let $(z_n : n \in \mathbf{N})$ denote an enumeration of $\mathbf{Q}^p$.

If G is an arbitrary open set in $\mathbf{R}^p$, for each $z \in G \cap \mathbf{Q}^p$, let $C(z, 1/n_z)$ be the open cube (i.e., the cell all of whose sides are equal) with center z and side length $1/n_z$, where n_z is the smallest natural number such that $C(z, 1/n_z)$ is contained in G. (We have just used the Well-ordering Property of N.) We now let

$$G_0 := \bigcup \{ C(z, 1/n_z) : z \in G \cap Q^p \}$$

so that G_0 is a union of a countable collection of open cubes. Therefore G_0 is open and $G_0 \subseteq G$.

We now show that $G_0 = G$. Indeed, if $y \in G$ is arbitrary, then since G is open, there exists a cube $C(y, 1/n_y)$ with center y such that $C(y, 1/n_y) \subseteq G$. Now consider the cube $C(y, 1/2n_y)$ whose side length is half that of $C(y, 1/n_y)$. Since Q is dense in R, then the set Q^p is dense in R^p, and there are infinitely many "rational" points in $G \cap C(y, 1/2n_y)$. Let w be the first such rational point, according to the enumeration of Q^p selected above. Since $w \in G \cap C(w, 1/2n_y)$, it is an exercise to show that $y \in C(w, 1/2n_y) \subseteq C(w, 1/n_w) \subseteq G_0$.

We conclude that $G = G_0$, so that G is the union of a countable collection of open cubes, which are open cells.   Q.E.D.

14.2   THEOREM.   *Every open and every closed subset of R^p is Lebesgue measurable.*

PROOF. It follows from Lemma 14.1 and the fact that the collection $\mathcal{L}$ possesses property 13.1(iii), that every open set belongs to $\mathcal{L}$. Since a closed subset of R^p is (by definition) the complement of an open set in R^p, every closed set is also Lebesgue measurable. Q.E.D.

The intersection of a sequence (or countable collection) of open sets is often called a G_δ-set. It is well known that such a set is not necessarily an open set (give an example!); however, it is a measurable set, since the intersection of a sequence of sets in a σ-algebra also belongs to the σ-algebra. Similarly, the union of a sequence (or countable collection) of closed sets is often called an F_σ-set. Such a set is not necessarily a closed set (example, please), but it is a measurable set. Further, a set that is the union of a sequence of G_δ-sets is often called a $G_{\delta\sigma}$-set; it is always a measurable set. Also, a set that is the intersection of a sequence of F_σ-sets is often called an $F_{\sigma\delta}$-set; it is also measurable. If we continue in this way, we can define sets that are

$$G_{\delta\sigma\delta}, \quad F_{\sigma\delta\sigma}, \quad G_{\delta\sigma\delta\sigma}, \quad F_{\sigma\delta\sigma\delta}, \quad \ldots$$

sets. All the sets that are obtained in this way are Lebesgue measurable sets, so it is very difficult even to imagine a subset of R^p that is not Lebesgue measurable. However, we will give an example in Chapter 17 of a subset of R^p that is *not* Lebesgue measurable.

BOREL SETS

In addition to the σ-algebra $\mathcal{L}$ of Lebesgue measurable sets, it is often convenient to work with a somewhat smaller σ-algebra of subsets of R^p.

14.3 DEFINITION. The smallest σ-algebra of subsets of R^p that contains all of the open sets is called the **Borel σ-algebra** and will be denoted by $\mathcal{B}$. Any set in $\mathcal{B}$ is called a **Borel set**.

A word needs to be said about the existence of the "smallest" σ-algebra containing the open sets. As we have noted in Remark (e) after Definition 13.1, the intersection of two σ-algebras is again a σ-algebra; moreover, it is an exercise to show that the intersection of an *arbitrary* collection of σ-algebras is a σ-algebra. Consequently, it follows from this observation that the intersection of all σ-algebras containing the open sets in R^p is a σ-algebra that contains the open sets; hence it is precisely the collection of Borel sets.

In addition to all open sets, and all closed sets, the collection $\mathcal{B}$ must also contain all G_δ-, all F_σ-, all $G_{\delta\sigma}$-, all $F_{\sigma\delta}$-, ... sets. On the other hand, since $\mathcal{L}$ is a σ-algebra containing all open sets, it follows that we must have

$$\mathcal{B} \subseteq \mathcal{L}.$$

The question naturally arises as to whether we might have equality in this inclusion; that is, whether every Lebesgue measurable set is a Borel set. The answer is: No. We will sketch a proof of this assertion at the end of this chapter.

NULL SETS

We now introduce a class of sets that are small (at least from the point of view of measure theory) but which often play a very important role.

14.4 DEFINITION.   A subset $E \subseteq R^p$ with $m^*(E) = 0$ is called a (Lebesgue) **null set**.

Of course, the empty set $\emptyset$ is a null set, as is a set consisting of a finite number of points. Indeed, any countable subset $Z :=$ $\{p_1, p_2, \ldots\}$ of R^p is a null set; for, given any $\varepsilon > 0$, let the point p_1 be enclosed in a cell I_1 with $m(I_1) < \varepsilon/2$, let p_2 be enclosed in a cell I_2 with $m(I_2) < \varepsilon/2^2, \ldots$, let p_n be enclosed in a cell I_n with $m(I_n) < \varepsilon/2^n, \ldots$. Hence, it follows from the countable subadditivity of m (which is a consequence of this same property of m^*) that

$$0 \leq m(Z) \leq \sum_{n=1}^{+\infty} m(I_n) \leq \sum_{n=1}^{+\infty} \varepsilon/2^n = \varepsilon.$$

We now show that any null set, and hence any subset of a null set, is a Lebesgue measurable set.

14.5 THEOREM.   *If $Z \subseteq R^p$ is a null set, then Z is a Lebesgue measurable set and hence $m(Z) = 0$. Moreover, any subset of Z is Lebesgue measurable and a null set.*

PROOF.   Let $A \subseteq R^p$ be arbitrary. Then since $Z \supseteq A \cap Z$ and $A \supseteq A \cap Z^c$, it follows from the monotone property of m^* that

$$m^*(A) = m^*(Z) + m^*(A) \geq m^*(A \cap Z) + m^*(A \cap Z^c).$$

Therefore, it is a consequence of Lemma 13.4 that Z is Lebesgue measurable; hence $m(Z) = m^*(Z) = 0$.

If $W \subseteq Z$, then $0 \leq m^*(W) \leq m^*(Z) = 0$, whence W is also a null set, and therefore is Lebesgue measurable.   Q.E.D.

One sometimes says that Lebesgue measure is **complete**, meaning that any subset of a Lebesgue null set is measurable, and hence has Lebesgue measure equal to 0.

Normally, we think of null sets as not having many points; however, that is not necessarily the case, as we will now show.

14.6 COROLLARY.   *There are Lebesgue null subsets of R^p that contain uncountably many points.*

PROOF. We will prove the assertion for the case $p = 1$, and leave it to the reader to extend the result to an arbitrary value of p.

We note that the familiar Cantor set [see pp. 351–353, *Introduction to Real Analysis*, Second edition, John Wiley & Sons, 1992, by R. G. Bartle and D. R. Sherbert] is a closed subset of $\boldsymbol{R}$. Further, the argument presented there shows that, given $\varepsilon > 0$, the Cantor set can be enclosed in the union of countably many cells whose total length is less than ε. Therefore, the Cantor set is a Lebesgue null set. However, it was also seen in the reference cited that the Cantor set contains uncountably many points.   Q.E.D.

TRANSLATION INVARIANCE

The next result can be summarized by saying "Lebesgue measure is translation invariant".

14.7 THEOREM. *If $E \subseteq \boldsymbol{R}^p$ is Lebesgue measurable and $x \in \boldsymbol{R}^p$, then $x \oplus E$ is Lebesgue measurable and*

$$m(x \oplus E) = m(E).$$

PROOF. If A and B are arbitrary subsets of $\boldsymbol{R}^p$ and if $z \in \boldsymbol{R}^p$, then it is an exercise to show that $y \in (z \oplus A) \cap B$ if and only if $y \in z \oplus (A \cap [(-z) \oplus B])$. Hence we have

$$(z \oplus A) \cap B = z \oplus \left(A \cap [(-z) \oplus B] \right).$$

Similarly, one shows that $x \oplus B^c = (x \oplus B)^c$. Now let x, z be related by $x = -z$; then it follows from the invariance of m^* under translation that

(14.1) $m^* \left((z \oplus A) \cap B \right) = m^* \left(A \cap (x \oplus B) \right).$

Now let $E \in \mathcal{L}$ and use (14.1) with $B = E$ and $B = E^c$ to obtain

$$\begin{aligned}
m^*(A) = m^*(z \oplus A) &= m^* \left((z \oplus A) \cap E \right) + m^* \left((z \oplus A) \cap E^c \right) \\
&= m^* \left(A \cap (x \oplus E) \right) + m^* \left(A \cap (x \oplus E^c) \right) \\
&= m^* \left(A \cap (x \oplus E) \right) + m^* \left(A \cap (x \oplus E)^c \right).
\end{aligned}$$

for all $A \subseteq \boldsymbol{R}^p$. Therefore $x \oplus E$ is also Lebesgue measurable and, from Theorem 12.6, it has Lebesgue measure equal to

$$m^*(x \oplus E) = m^*(E) = m(E). \qquad \text{Q.E.D.}$$

NON-BOREL SETS

We now return to the question of whether there exists a Lebesgue measurable set that is not a Borel set. We will state the result formally in the case that $p = 1$, but give only an outline of the argument, which is based on cardinal numbers. A reader not familiar with cardinal numbers will do well to accept the validity of the assertion on faith until Chapter 17, where a more "constructive" proof is given.

14.8 THEOREM. *In the space $\boldsymbol{R}$ there exist Lebesgue measurable sets that are not Borel sets.*

SKETCH OF PROOF. We first notice that there is a countable number of open cells with rational endpoints; that is, the cardinality of the collection of all of these "rational cells" is equal to the cardinal number $\aleph_0$. It is an exercise to show that $\mathcal{B}$ is the smallest σ-algebra containing all of these "rational cells". Therefore, it follows that the cardinal number of the set $\mathcal{B}$ is

$$(\aleph_0)^{\aleph_0} = c,$$

where c denotes the cardinality of the set $\boldsymbol{R}$ of all real numbers. On the other hand, it was seen in Corollary 14.6 that $\mathcal{L}$ contains a null set with uncountably many elements; in fact, the Cantor set can be seen to have c elements. Since an arbitrary subset of a null set is a Lebesgue measurable set, then $\mathcal{L}$ contains at least 2^c sets. But since $\boldsymbol{R}$ has only 2^c subsets, it follows that the cardinality $\text{card}(\mathcal{L}) = 2^c$. Therefore, we have

$$\text{card}(\mathcal{B}) = c < 2^c = \text{card}(\mathcal{L}),$$

whence $\mathcal{B}$ is properly contained in $\mathcal{L}$.

CHAPTER 15

Approximation of Measurable Sets

We now show that arbitrary Lebesgue measurable sets are precisely those sets that can be approximated by open and closed sets. In fact, it would be possible to take these characterizations as the definition of measurability. However, we have chosen not to do so, since in abstract measure spaces there is no notion of "open" or "closed", so the Carathéodory condition is the only method that is possible of determining measurability.

APPROXIMATION BY OPEN SETS

First we will show that every subset of R^p can be enclosed in a G_δ-set with the same outer measure.

15.1 LEMMA. (a) *If $A \subseteq R^p$ and $\varepsilon > 0$, then there exists an open set $G \subseteq R^p$ such that $A \subseteq G$ and $m(G) \leq m^*(A) + \varepsilon$. Hence*

(15.1) $$m^*(A) = \inf\{m(G) : A \subseteq G, \ G \text{ open}\}.$$

(b) *If $A \subseteq R^p$, then there is a G_δ-set H such that $A \subseteq H$ and $m^*(A) = m(H)$.*

PROOF. (a) We may assume that $m^*(A) < +\infty$. (Why?) It was noted in Remark 12.2(b) that we may require the cells in Definition

155

12.1 to be open. Thus there exists a sequence (I_k) of open cells covering the set A such that

$$\sum_{k=1}^{+\infty} l(I_k) \leq m^*(A) + \varepsilon.$$

If we let $G := \bigcup_{k=1}^{+\infty} I_k$, then G is open and, by the countable subadditivity of m^* and Theorem 12.5, we have that

$$m(G) \leq \sum_{k=1}^{+\infty} m^*(I_k) = \sum_{k=1}^{+\infty} l(I_k) \leq m^*(A) + \varepsilon.$$

Equation (15.1) now follows from the definition of the infimum.

(b) For each $n \in N$, let G_n be an open set such that $A \subseteq G_n$ and $m(G_n) \leq m^*(A) + 1/n$. Now let $H := \bigcap_{n=1}^{+\infty} G_n$, so that $A \subseteq H \subseteq G_n$ and $m^*(A) \leq m(H) \leq m^*(A) + 1/n$ for all $n \in N$. Therefore $m^*(A) = m(H)$, as asserted.   Q.E.D.

15.2 COROLLARY. *Every Lebesgue null set is a subset of a Borel null set.*

PROOF. If Z is a Lebesgue null set, there is a G_δ-set H such that $Z \subseteq H$ and $m(H) = 0$. But H is a Borel set.   Q.E.D.

Unfortunately, in Lemma 15.1(b), the difference $H - A$ need not be a "small" set. In fact, it will be seen in Corollary 15.5 that the set A is Lebesgue measurable if and only if the set $H - A$ is a null set.

15.3 THEOREM. *A set $E \subseteq R^p$ is Lebesgue measurable if and only if for every $\varepsilon > 0$ there exists an open set G with $E \subseteq G$ and $m^*(G - E) < \varepsilon$.*

PROOF. We first assume that E is measurable and that $m(E) < +\infty$. Then, by Lemma 15.1(a), there exists an open set G such that $E \subseteq G$ and $m(G) < m(E) + \varepsilon$. Since E is measurable and $E \subseteq G$, we have

$$m(G) = m(G \cap E) + m(G - E) = m(E) + m(G - E).$$

Since $m(E) < +\infty$, we have

$$m(G - E) = m(G) - m(E) < \varepsilon.$$

If $m(E) = +\infty$, let $E_1 := E \cap \{x : \|x\| \le 1\}$ and, if $n \ge 2$, let $E_n := E \cap \{x : n - 1 < \|x\| \le n\}$. For $n \in N$, let G_n be an open set with $E_n \subseteq G_n$ and $m(G_n - E_n) < \varepsilon/2^n$. If we let $G := \bigcup_{n=1}^{+\infty} G_n$, then G is open, $E \subseteq G$, and

$$G - E \subseteq \bigcup_{n=1}^{+\infty}(G_n - E_n).$$

Therefore, from the countable subadditivity of m^*, we have

$$m(G - E) \le \sum_{n=1}^{+\infty} m(G_n - E_n) < \sum_{n=1}^{+\infty} \varepsilon/2^n = \varepsilon.$$

Conversely, suppose that for every $n \in N$ there exists an open set $G_n \supseteq E$ such that $m^*(G_n - E) < 1/n$. Let $H := \bigcap_{n=1}^{+\infty} G_n$ so that H is a G_δ-set (and hence is measurable). Moreover, since $H \subseteq G_n$, we have $H - E \subseteq G_n - E$ and hence

$$0 \le m^*(H - E) \le m^*(G_n - E) < 1/n$$

for all $n \in N$. Therefore $m^*(H - E) = 0$, which implies that $Z := H - E$ is a measurable set. Therefore, $E = H - Z$ is a measurable set.   Q.E.D.

15.4 COROLLARY. *If $E \subseteq R^p$ is measurable, then for any $\varepsilon > 0$ there exists an open set $G \supseteq E$ with $m(G) \le m(E) + \varepsilon$. Therefore, we have*

$$m(E) = \inf\{m(G) : G \text{ open}, \ G \supseteq E\}.$$

PROOF. Indeed, from Theorem 15.3, we have $m(G) = m(E) + m(G - E) \le m(E) + \varepsilon$.   Q.E.D.

The next result is concerned with approximation of a set by a G_δ-set from the outside. It is a useful characterization of Lebesgue measurability.

15.5 COROLLARY. *The following statements are equivalent:*

(a) *The set $E \subseteq \boldsymbol{R}^p$ is Lebesgue measurable;*

(b) *there exists a G_δ-set H with $E \subseteq H$ and $m^*(H - E) = 0$;*

(c) *there exist a G_δ-set H and a Lebesgue null set Z such that $E \subseteq H$, $Z \subseteq H$, and $E = H - Z$.*

We leave it as an exercise to write out the details of the proof.

APPROXIMATION BY CLOSED SETS

We now show that Lebesgue measurable sets are those sets that can be well approximated from inside by closed sets.

15.6 THEOREM. *A set $E \subseteq \boldsymbol{R}^p$ is Lebesgue measurable if and only if for each $\varepsilon > 0$ there exists a closed set F with $F \subseteq E$ and $m^*(E - F) < \varepsilon$.*

PROOF. If E is measurable, then its complement E^c is also measurable. By Theorem 15.3 there exists an open set G with $E^c \subseteq G$ and $m(G - E^c) < \varepsilon$. Now let $F := G^c$ so that F is closed, $F \subseteq E$ and $E - F = E \cap G = G - E^c$, whence $m(E - F) = m(G - E^c) < \varepsilon$.

Conversely, for each $n \in \boldsymbol{N}$ there is a closed set $F_n \subseteq E$ with $m^*(E - F_n) < 1/n$, and we let $K := \bigcup_{n=1}^{+\infty} F_n$. Then K is a F_σ-set (and is therefore Lebesgue measurable) and, since $F_n \subseteq K$, we have $E - K \subseteq E - F_n$, so that $m^*(E - K) \leq m^*(E - F_n) < 1/n$ for all n. Therefore $m^*(E - K) = 0$ which implies that $Z := E - K$ is a measurable set. Therefore $E = K \cup Z$ is measurable.   Q.E.D.

15.7 COROLLARY. *If $E \subseteq \boldsymbol{R}^p$ is Lebesgue measurable, then for any $\varepsilon > 0$ there exists a closed set $F \subseteq E$ with $m(E) \leq m(F) + \varepsilon$. Therefore, we have*

$$m(E) = \sup\{m(F) : F \text{ closed}, F \subseteq E\}.$$

PROOF. The set F in the theorem is measurable and $F \subseteq E$; hence $m(E) = m^*(E \cap F) + m^*(E - F) \leq m(F) + \varepsilon$.   Q.E.D.

The next corollary characterizes measurability in terms of the approximation by F_σ-sets from inside.

15.8 COROLLARY. *The following statements are equivalent:*

(a) *The set $E \subseteq R^p$ is Lebesgue measurable;*

(b) *there exists an F_σ-set K with $K \subseteq E$ and $m^*(E-K) = 0$;*

(c) *there exist an F_σ-set K and a Lebesgue null set Z such that $K \subseteq E$, $Z \subseteq E$, and $E = K \cup Z$.*

We leave it as an exercise to write out the details of the proof.

APPROXIMATION BY COMPACT SETS

If E is a Lebesgue measurable set with $m(E) < +\infty$, then we can approximate it from within by *compact* sets, and conversely. We recall that a compact set always has finite Lebesgue measure.

15.9 THEOREM. *A set $E \subseteq R^p$ with $m^*(E) < +\infty$ is Lebesgue measurable if and only if, for every $\varepsilon > 0$ there exists a compact set C with $C \subseteq E$ and $m^*(E - C) < \varepsilon$.*

PROOF. If E is measurable and $n \in N$, let E_n be the set defined by $E_n := E \cap \{x : \|x\| \leq n\}$. Since the sequence (E_n) increases to E, it follows that the numerical sequence $(m(E_n))$ also increases to $m(E) < +\infty$, so there is an n_0 such that $m(E) < m(E_{n_0}) + \varepsilon/2$. By Theorem 15.6 there is a closed set C with $C \subseteq E_{n_0}$ and $m(E_{n_0} - C) < \varepsilon/2$. Since E is the union of the disjoint sets $E - E_{n_0}$ and E_{n_0}, it follows that

$$m(E) = m(E - E_{n_0}) + m(E_{n_0})$$

and since $m(E) < +\infty$, that

$$m(E - E_{n_0}) = m(E) - m(E_{n_0}) < \varepsilon/2.$$

In addition, $E - C$ is the union of the disjoint sets $E - E_{n_0}$ and $E_{n_0} - C$. Therefore

$$m(E - C) = m(E - E_{n_0}) + m(E_{n_0} - C) < \varepsilon.$$

Since $C \subseteq E_{n_0}$ is closed and bounded, it is compact in R^p.

Conversely, suppose that for every $n \in N$ there is a compact set C_n with $C_n \subseteq E$ and $m^*(E - C_n) < 1/n$. If we set $C := \bigcup_{n=1}^{+\infty} C_n$, then C is measurable and $E - C \subseteq E - C_n$ for all natural numbers $n \in \mathbf{N}$, and it follows that $m^*(E-C) = 0$. Therefore $Z := E - C$ is a Lebesgue null set and hence is measurable. Consequently, $E = C \cup Z$ is Lebesgue measurable.   Q.E.D.

APPROXIMATION BY CELLS

We now show that a set with finite measure can be approximated by a finite union of bounded cells. We recall that the **symmetric difference** of two sets A, B is the set $A \triangle B := (A - B) \cup (B - A)$.

15.10 THEOREM.   *If $E \in \mathcal{L}$ has finite measure and $\varepsilon > 0$, then there exist bounded open cells $I_1, \ldots, I_n$ such that if $K := \bigcup_{i=1}^{n} I_i$, then $m(E \triangle K) < \varepsilon$.*

PROOF.   As in the proof of Lemma 15.1, there exists a sequence $(I_i)_{i=1}^{+\infty}$ of bounded open cells covering E such that if $G := \bigcup_{i=1}^{+\infty} I_i$, then $m(G) \le m(E) + \varepsilon/2$. Similarly, by Theorem 15.9, there exists a compact set $C \subseteq E$ such that $m(E - C) < \varepsilon/2$. It follows from the Heine-Borel Theorem that a finite number of the cells, say $I_1, \ldots, I_n$ cover C. If $K := \bigcup_{i=1}^{n} I_i$, then since $C \subseteq K \subseteq G$ and $C \subseteq E \subseteq G$, it follows that

$$m(E \triangle K) = m(E - K) + m(K - E)$$
$$\le m(E - C) + m(G - E) < \varepsilon. \qquad \text{Q.E.D.}$$

It is left as an exercise to show that the cells $I_1, \ldots, I_n$ can be chosen to be closed, or half-open, or pairwise disjoint.

CHAPTER 16

Additivity and Nonadditivity

In this brief chapter we establish the surprising fact that the outer measure function m^* is additive over the union of two disjoint sets provided that *one* of them is measurable. We will also give some other results concerning the additivity and nonadditivity properties of m^*. In addition, we will show that if a set is known to be contained in a measurable set with finite measure, then the Carathéodory condition can be replaced by a *single* test, and relate this result with the notion of "inner measure".

16.1 THEOREM. *Let E be a Lebesgue measurable subset of $\mathbf{R}^p$ and let F be any subset of $\mathbf{R}^p$. Then:*

(a) $m^*(E \cup F) + m^*(E \cap F) = m(E) + m^*(F)$;

(b) *if $E \cap F = \emptyset$, then $m^*(E \cup F) = m(E) + m^*(F)$;*

(c) *if $m(E) < +\infty$ and $E \subseteq F$, then we have $m^*(F - E) = m^*(F) - m(E)$.*

PROOF. Since $E \in \mathcal{L}$, it follows from the Carathéodory condition that $m^*(A) = m^*(A \cap E) + m^*(A - E)$ for all $A \subseteq \mathbf{R}^p$. If we take $A := E \cup F$, we obtain

$$m^*(E \cup F) = m^*\big((E \cup F) \cap E\big) + m^*\big((E \cup F) - E\big)$$
$$= m(E) + m^*(F - E),$$

and if we take $A := F$, we obtain

$$m^*(F) = m^*(F \cap E) + m^*(F - E).$$

Therefore, we have

$$
\begin{aligned}
m^*(E \cup F) + m^*(E \cap F) &= \big[m(E) + m^*(F - E)\big] + m^*(E \cap F) \\
&= m(E) + \big[m^*(F - E) + m^*(E \cap F)\big] \\
&= m(E) + m^*(F),
\end{aligned}
$$

as asserted.

(b) If $E \cap F = \emptyset$, then $m^*(E \cap F) = 0$, so the conclusion is immediate.

(c) Let $H := F - E$, so that $F = E \cup H$ and $E \cap H = \emptyset$. Hence, from (b) we have

$$
\begin{aligned}
m^*(F) = m^*(E \cup H) &= m(E) + m^*(H) \\
&= m(E) + m^*(F - E).
\end{aligned}
$$

Since $m(E) < +\infty$, we know that $m^*(F)$ and $m^*(F - E)$ are either both $+\infty$ or both finite. Hence the assertion follows.   Q.E.D.

We saw in Lemma 15.1(b) that for any set $E \subseteq \mathbf{R}^p$ there exists a G_δ-set H such that $E \subseteq H$ and $m^*(E) = m(H)$. It also follows that E is Lebesgue measurable if and only if $H - E$ is a null set. The next theorem is somewhat remarkable in that a similar approximation from within guarantees the measurability of a set E with finite outer measure merely from the equality of the outer measures of the sets.

16.2 THEOREM. *If $m^*(E) < +\infty$, then E is measurable if and only if there is a measurable set $B \subseteq E$ with $m(B) = m^*(E)$.*

PROOF. If E is measurable, then we can obviously take $B := E$.

On the other hand, if $B \in \mathcal{L}$, $B \subseteq E$, and $m(B) = m^*(E)$, then it follows from Theorem 16.1(c) that

$$m^*(E - B) = m^*(E) - m(B) = 0.$$

Consequently, the null set $E-B$ is Lebesgue measurable and therefore $E = (E - B) \cup B$ is also measurable.   Q.E.D.

CARATHÉODORY REVISITED

We now obtain a modification of the Carathéodory condition that involves only testing with a single measurable set of finite measure that contains the given set.

16.3 THEOREM.   *Let $A \subseteq R^p$ be Lebesgue measurable with $m(A) < +\infty$. Then $E \subseteq A$ is Lebesgue measurable if and only if*

$$(16.1) \qquad\qquad m(A) = m^*(E) + m^*(A - E).$$

Proof. If E is measurable, the assertion follows immediately from the Carathéodory condition.

Conversely, by Theorem 15.1(b) applied to the set $A - E$, there is a G_δ-set H with $A - E \subseteq H$ and $m^*(A - E) = m(H)$. Since $A - E \subseteq A \cap H \subseteq H$, it follows that

$$m^*(A - E) \le m(A \cap H) \le m(H) = m^*(A - E),$$

whence $m(A \cap H) = m^*(A - E)$. But since $A \cap H$ is measurable and

$$A \cap (A \cap H) = A \cap H \qquad \text{and} \qquad A - (A \cap H) = A - H,$$

we conclude that

$$m(A) = m(A \cap H) + m(A - H) = m^*(A - E) + m(A - H).$$

If we use equation (16.1), we deduce that

$$m(A - H) = m^*(E).$$

But since $B := A - H \subseteq E$, it follows from Theorem 16.2 that E is Lebesgue measurable, as claimed.   Q.E.D.

It is often the case that we are concerned with sets that are contained in some large cell, such as

$$J_n := [-n, n] \times \ldots \times [-n, n],$$

for some $n \in N$. The preceding theorem is useful in establishing the measurability of a set $E \subseteq J_n$.

16.4 COROLLARY. *A set $E \subseteq J_n$ is Lebesgue measurable if and only if*

$$(16.2) \qquad m(J_n) = m^*(E) + m^*(J_n - E).$$

PROOF. This follows immediately from Theorem 16.3 and the fact that J_n is measurable (Theorem 13.7).   Q.E.D.

For an unbounded set $E \subseteq R^p$, the next result is useful.

16.5 THEOREM. *A set $E \subseteq R^p$ is Lebesgue measurable if and only if the sets $E \cap J_n$ are measurable for each $n \in N$.*

PROOF. If E is measurable, then the result is trivial.

Conversely, if each set $E_n := E \cap J_n$ is measurable, then it follows from the fact that $E = \bigcup_{n=1}^{+\infty} E_n$, that E is measurable.   Q.E.D.

INNER MEASURE

Readers may initially have been surprised that we have focused our attention on outer measure and have not defined a notion of the "inner measure" of a set by *inscribing* a collection of cells inside a given set and taking the supremum of the resulting inner approximations. One of the reasons we have not done so is that, in general, a measurable set — even one with positive measure — may not contain any cells with positive volume. (Construct an example, please.) However, there is a way around this difficulty for a set E that is contained in a cell J_n. Namely, we define the **inner measure** $m_*(E)$ of E to be the difference

$$m_*(E) := m(J_n) - m^*(J_n - E).$$

With this definition, Corollary 16.4 takes the form: *A set $E \subseteq J_n$ is Lebesgue measurable if and only if its inner measure and its outer measure are equal.* This observation is often used in the process of defining Lebesgue measure in the interval $[0, 1]$.

CHAPTER 17

Nonmeasurable and Non-Borel sets

In this chapter we will establish the existence of a set in R^p that is not measurable. In order to do so, we will need to use the "Axiom of Choice". The first person to give an example of a nonmeasurable set was Giuseppe Vitali, in 1905. In 1970, R. M. Solovay showed that the use of the Axiom of Choice is essential, in a certain technical sense. Although other constructions of nonmeasurable sets have been given, the author is not aware of any that are simpler than Vitali's example. We will also obtain some other results about the nonadditivity that is characteristic of nonmeasurable sets, and establish the existence of a nonmeasurable set such that neither the set nor its complement contains any measurable sets that are not null sets. At the end of this chapter we give a proof of the existence of a Lebesgue measurable set in R that is not a Borel set.

The following definition will be useful.

17.1 DEFINITION. If $A \subseteq R^p$, then its **difference set** is defined to be

$$A \ominus A := \{x - y : x, y \in A\}.$$

It is trivial, but useful, to observe that if $A \subseteq B$, then $A \ominus A \subseteq B \ominus B$.

165

17.2 LEMMA.   *Let $K \subseteq \mathbf{R}^p$ be a compact set with $m(K) > 0$. Then the difference set $K \ominus K$ contains an open ball with center at the origin of $\mathbf{R}^p$.*

PROOF.   Since $0 < m(K) < +\infty$, there exists an open set G such that $K \subseteq G$ and $m(G) < 2m(K)$. Since K is compact and $G^c = \mathbf{R}^p - G$ is closed, we conclude that

$$\delta := \operatorname{dist}(K, G^c) > 0.$$

This implies that if $\|x\| = \operatorname{dist}(x, 0) < \delta$, then $x \oplus K \subseteq G$.

We claim that $(x \oplus K) \cap K \neq \emptyset$. For, if not, then since we have $K \cup (x \oplus K) \subseteq G$, it follows from $(x \oplus K) \cap K = \emptyset$ and the additivity of m that

$$2m(K) = m(K) + m(x \oplus K) = m(K \cup (x \oplus K))$$
$$\leq m(G) < 2m(K),$$

which is a contradiction. Therefore $(x \oplus K) \cap K \neq \emptyset$ for all x with $\|x\| < \delta$. But this implies that if $\|x\| < \delta$, then there exist $k_1, k_2 \in K$ such that $x = k_1 - k_2 \in K \ominus K$. Therefore the set $K \ominus K$ contains the open ball with center 0 and radius δ.   Q.E.D.

17.3   THEOREM.   *If $E \subseteq \mathbf{R}^p$ is any Lebesgue measurable set with $m(E) > 0$, then the difference set $E \ominus E$ contains an open ball with center 0.*

PROOF.   For $n \in \mathbf{N}$, let $E_n := \{x \in E : \|x\| < n\}$. Since $m(E) = \lim_n m(E_n)$, we have $m(E_n) > 0$ for n sufficiently large, say for all $n \geq n_0$. We note that $0 < m(E_{n_0}) < +\infty$. By Theorem 15.9 there exists a compact set $K \subseteq E_{n_0} \subseteq E$ with $0 < (1/2)m(E_{n_0}) \leq m(K)$. Since $K \subseteq E$, it is clear that $K \ominus K \subseteq E \ominus E$. By the preceding lemma, we conclude that $K \ominus K$ contains an open ball with center 0; therefore, so does $E \ominus E$.   Q.E.D.

17.4   DEFINITION.   If $x, y \in \mathbf{R}^p$, we say that x is **rationally equivalent** to y and write $x \sim y$ if $x - y \in \mathbf{Q}^p$; that is, if the components $x_i - y_i \in \mathbf{Q}$ for all $i = 1, 2, \ldots, p$.

It is easy to see that rational equivalence is an equivalence relation on the set $\boldsymbol{R}^p$ [in the sense that (i) $x \sim x$; (ii) $x \sim y$ if and only if $y \sim x$; and (iii) $x \sim y$ and $y \sim z$ imply that $x \sim z$]. Hence rational equivalence on $\boldsymbol{R}^p$ divides $\boldsymbol{R}^p$ into a collection of disjoint equivalence classes. Using the Axiom of Choice, we form a set $\mathcal{V}$ by choosing one representative from each equivalence class, so that if $v_1, v_2 \in \mathcal{V}$ and $v_1 \sim v_2$, then $v_1 = v_2$. We refer to any such set $\mathcal{V}$ as **Vitali's set**. Moreover, if $q \in \boldsymbol{Q}^p$, we write

$$(17.1) \qquad \qquad \mathcal{V}_q := q \oplus \mathcal{V}.$$

It is easy to see that if $q, q' \in \boldsymbol{Q}^p, q \neq q'$, then $\mathcal{V}_q \cap \mathcal{V}_{q'} = \emptyset$.

17.5 LEMMA. *Let $q_1, q_2, \ldots$ be an enumeration of the countable set $\boldsymbol{Q}^p$. Then $\boldsymbol{R}^p$ can be represented by the disjoint union:*

$$(17.2) \qquad \qquad \boldsymbol{R}^p = \bigcup_{i=1}^{+\infty} \mathcal{V}_{q_i} = \bigcup_{i=1}^{+\infty} (q_i \oplus \mathcal{V}).$$

Therefore, every element $x \in \boldsymbol{R}^p$ has a unique representation in the form $x = q_i + v$ for some q_i and some $v \in \mathcal{V}$.

PROOF. If $x \in \boldsymbol{R}^p$, then x belongs to a unique rational equivalence class. If v is the representative of this class, then $x - v = q_i$ for some i, so $x = q_i + v \in \mathcal{V}_{q_i}$. If $i \neq j$, then the sets $\mathcal{V}_{q_i}$ and $\mathcal{V}_{q_j}$ are disjoint, as noted above.   Q.E.D.

17.6 THEOREM. *Vitali's set $\mathcal{V}$ is not Lebesgue measurable.*

PROOF. Suppose, on the contrary, that the set $\mathcal{V}$ is measurable. We have two possibilities: (i) $m(\mathcal{V}) > 0$, or (ii) $m(\mathcal{V}) = 0$.

Case (i).   If $m(\mathcal{V}) > 0$, then Theorem 17.3 implies that the difference set $\mathcal{V} \ominus \mathcal{V}$ contains an open ball with center 0. Therefore there exists a nonzero element x in this ball, all of whose coordinates are rational numbers. Since x belongs to this ball, there are elements $v_1, v_2 \in \mathcal{V}$ such that $x = v_1 - v_2$. But this means that $v_1 \sim v_2$, whence we conclude that $v_1 = v_2$ and $x = 0$, a contradiction. Therefore, this case is not possible.

Case (ii).   Suppose that $m(\mathcal{V}) = 0$. Since Lebesgue measure is translation invariant, we have $m(q \oplus \mathcal{V}) = 0$ for all $q \in \mathbf{Q}^p$. It follows from the countable additivity of m and from (17.2) that

$$0 \le m(\mathbf{R}^p) = \sum_{i=1}^{+\infty} m(q_i \oplus \mathcal{V}) = 0.$$

Therefore, we have $m(\mathbf{R}^p) = 0$, which implies that the measure of every measurable subset in $\mathbf{R}^p$ is equal to 0, which is a contradiction.

Therefore, Vitali's set is not Lebesgue measurable.   Q.E.D.

We now show that *any* set with positive outer measure contains a subset that is not Lebesgue measurable. (Thus, nonmeasurable sets are everywhere in $\mathbf{R}^p$!)

17.7   THEOREM.   *Any set $E \subseteq \mathbf{R}^p$ with $m^*(E) > 0$ contains a nonmeasurable subset.*

PROOF. By the preceding theorem, the translates $\mathcal{V}_{q_i} = q_i \oplus \mathcal{V}$ are not measurable; however, it is conceivable that their intersections $E_i := E \cap \mathcal{V}_{q_i}$ might be measurable. However, if E_i is measurable for some i and has positive measure, then it follows from Theorem 17.3 that the difference set $E_i \ominus E_i$ must contain a ball. But since $E_i \subseteq \mathcal{V}_{q_i}$, it follows that the difference set $\mathcal{V}_{q_i} \ominus \mathcal{V}_{q_i} = \mathcal{V} \ominus \mathcal{V}$ must also contain a ball, which is contrary to the construction of $\mathcal{V}$. We conclude that the sets E_i that are measurable must be null sets. It follows from (17.2) that

$$E = \bigcup_{i=1}^{+\infty} E \cap \mathcal{V}_{q_i} = \bigcup_{i=1}^{+\infty} E_i.$$

If all of the sets E_i are measurable, we have just seen that they must be null sets, so E is also a null set, contrary to hypothesis. Therefore, at least one of the sets E_i is not Lebesgue measurable.   Q.E.D.

We now show that every measurable set with finite positive measure has a nonadditive decomposition into the union of two sets that are nonmeasurable.

17.8 THEOREM. *Let E be a Lebesgue measurable set such that $0 < m(E) < +\infty$. Then there exist nonmeasurable subsets B and C of E such that $E = B \cup C$, $B \cap C = \emptyset$, and*

$$(17.3) \qquad m(E) < m^*(B) + m^*(C).$$

PROOF. It follows from Theorem 17.7 that the set E has a non-measurable subset B, and we let $C := E - B$, so that $E = B \cup C$ and $B \cap C = \emptyset$. Moreover, since $B = E - C$, the set C must also be nonmeasurable. We conclude from the subadditivity of m^* that

$$(17.4) \qquad m^*(E) \leq m^*(B) + m^*(C),$$

However, if equality holds in (17.4), then it follows from Theorem 16.3 that B is a measurable set, which is a contradiction. Therefore, we have (17.3). Q.E.D.

We will now see that every nonmeasurable set with finite outer measure is part of a nonadditive decomposition of a measurable set.

17.9 THEOREM. *Let B be a nonmeasurable set such that $m^*(B) < +\infty$, and let H be a G_δ-set set with $B \subseteq H$ and $m^*(B) = m(H)$. Then $C := H - B$ is also nonmeasurable and*

$$m(H) = m(B \cup C) < m^*(B) + m^*(C).$$

PROOF. The existence of H was established in Theorem 15.1(b). Since $B = H - C$, it follows that C must be nonmeasurable. Therefore $m^*(C) > 0$, whence the strict inequality follows. Q.E.D.

We now show that there exists a nonmeasurable set that can be said to be "ubiquitous", since every measurable subset of it (and of its complement) is a null set.

17.10 THEOREM. *There exists a nonmeasurable set $\mathcal{U} \subseteq \mathbf{R}^p$ such that every Lebesgue measurable subset of $\mathcal{U}$ is a null set, and every measurable subset of its complement $\mathcal{U}^c$ is also a null set.*

PROOF. Let G and G_0 be defined by

$$G := \mathbf{Q}^p \oplus \sqrt{2}\mathbf{Z}^p \quad \text{and} \quad G_0 := \mathbf{Q}^p \oplus (2\sqrt{2})\mathbf{Z}^p,$$

and let G_1 be defined by

$$G_1 := (\sqrt{2}, \dots, \sqrt{2}) \oplus G_0.$$

It is easy to see that

$$G = \{(q_1 + n_1\sqrt{2}, \dots, q_p + n_p\sqrt{2}) : q_i \in \mathbf{Q}, n_i \in \mathbf{Z}\},$$
$$G_0 = \{(q_1 + 2n_1\sqrt{2}, \dots, q_p + 2n_p\sqrt{2}) : q_i \in \mathbf{Q}, n_i \in \mathbf{Z}\},$$
$$G_1 = \{(q_1 + (2n_1 + 1)\sqrt{2}, \dots, q_p + (2n_p + 1)\sqrt{2}) : q_i \in \mathbf{Q}, n_i \in \mathbf{Z}\}.$$

We note that G and G_0 are subgroups of $\mathbf{R}^p$ under addition, and that G, G_0 and G_1 are dense in $\mathbf{R}^p$. It is also clear that

$$G = G_0 \oplus G_1 \quad \text{and} \quad G_0 \cap G_1 = \emptyset.$$

For each pair $x, y \in \mathbf{R}^p$, we define $x \approx y$ to mean that $x - y \in G$, so that $\approx$ is an equivalence relation on $\mathbf{R}^p$. We use the Axiom of Choice to obtain a set $\mathcal{E} \subseteq \mathbf{R}^p$ containing a single element from each equivalence class in $\mathbf{R}^p$, so that if $g \neq g'$, then $(g \oplus \mathcal{E}) \cap (g' \oplus \mathcal{E}) = \emptyset$. Consequently, we have the disjoint decomposition

$$\mathbf{R}^p = \bigcup_{g \in G} (g \oplus \mathcal{E}).$$

We now define our desired set by

(17.5) $$\mathcal{U} := G_0 \oplus \mathcal{E} = \bigcup_{g \in G_0} (g \oplus \mathcal{E}).$$

We note that since $G_0 \cap G_1 = \emptyset$, it follows that the complement $\mathcal{U}^c$ is given by

$$\mathcal{U}^c = G_1 \oplus \mathcal{E} = \bigcup_{g \in G_1} (g \oplus \mathcal{E}).$$

Let $F \subseteq \mathcal{U}$ be measurable; for the purpose of obtaining a contradiction, we assume that $m(F) > 0$. If so, then by Theorem 17.3 there is a ball contained in $F \ominus F \subseteq \mathcal{U} \ominus \mathcal{U}$. But, since G_1 is dense in $\mathbf{R}^p$, it must meet this ball and hence G_1 meets $\mathcal{U} \ominus \mathcal{U}$. In order for this to be true there must exist an element $g_1 \in G_1$ of the form $g_0 + e_1 - e_2$, where $g_0 \in G_0$ and $e_1, e_2 \in \mathcal{E}$. Therefore $e_1 - e_2 = g_1 - g_0 \in G$, which implies that $e_1 \approx e_2$ and hence $e_1 = e_2$ and $g_1 = g_0$. Since $G_1 \cap G_0 = \emptyset$, we have obtained a contradiction. Therefore, we conclude that any measurable subset of $\mathcal{U}$ is a null set.

We have already noted that $\mathcal{U}^c = G_1 \oplus \mathcal{E}$. Consequently, if $F_1 \subseteq \mathcal{U}^c$, then F_1 has the form

$$F_1 = (\sqrt{2}, \ldots, \sqrt{2}) \oplus F$$

for some subset F of $\mathcal{U}$. Moreover, F_1 is measurable if and only if its translate F is measurable. But it follows from the preceding paragraph that the only measurable subsets of $\mathcal{U}$ are null sets. By the translation invariance, we infer that F_1 is also a null set.

We conclude by noting that $\mathcal{U}$ must be nonmeasurable. For, if it is measurable, then its complement $\mathcal{U}^c = (\sqrt{2}, \ldots, \sqrt{2}) \oplus \mathcal{U}$, being a translate, is also measurable. By what we have seen above, both $\mathcal{U}$ and its complement $\mathcal{U}^c$ must be null sets. Consequently, $\mathbf{R}^p$ is a null set, a contradiction showing that $\mathcal{U}$ is nonmeasurable.   Q.E.D.

THE EXISTENCE OF NON-BOREL SETS

Let $\mathbf{F}$ be the Cantor set, obtained by deleting "middle thirds" from the unit interval $\mathbf{I} := [0, 1]$. If $x \in \mathbf{F}$, then x has a unique ternary (that is, base 3) expansion represented by

$$x = (0.c_1 c_2 c_3 \ldots)_3,$$

where $c_k = 0$ or 2 for all $k \in \mathbf{N}$. We define a mapping $\varphi : \mathbf{F} \to \mathbf{I}$ by

$$\varphi(x) := \big(0.(c_1/2)(c_2/2)(c_3/2) \ldots\big)_2,$$

using the binary (that is, base 2) expansion of the number. Clearly, if $x', x'' \in \mathbf{F}$ and $x' < x''$, then there exists $k \in \mathbf{N}$ such that all of the digits in the ternary expansion of x' are equal to those in the expansion of x'' up to the kth digit, but the kth digit in the ternary expansion of x' is 0 and the kth digit in the ternary expansion of x'' is 2. Hence it follows that $\varphi(x') \leq \varphi(x'')$ so that φ is a monotone nondecreasing map of F into I. Note, however, that φ is not one-one; for example, if $x' = (0.022\underline{0})_3 < x'' = (0.020\underline{2})_3$, then $\varphi(x') = (0.011\underline{0})_2 = (0.010\underline{1})_2 = \varphi(x'')$. In fact, $\varphi(x') = \varphi(x'')$ for $x' < x''$ if and only if these points have the form

$$x' = (0.c_1 c_2 \ldots c_k 0\underline{2})_3, \qquad x'' = (0.c_1 c_2 \ldots c_k 2\underline{0})_3,$$

and this holds if and only if x' and x'' are left and right end points of a middle third interval

$$(0.c_1 c_2 \ldots c_k 1\underline{0})_3 \leq x \leq (0.c_1 c_2 \ldots c_k 2\underline{0})_3,$$

removed in the process of constructing F. It is also clear that φ maps F onto I, since if $y = (0.b_1 b_2 \ldots)_2$ is the binary expansion of an arbitrary number in I, then y is the image under φ of the number $x = (0.(2b_1)(2b_2)\ldots)_3$ in F.

We now extend φ to be defined on all of I by defining it to be constant on each of the middle third sets that are removed from I in the construction of F. We have noted that φ takes the same value at both end points of such a middle third set. In particular,

$$\varphi(x) = (0.0\underline{1})_2 = (0.1\underline{0})_2 = \tfrac{1}{2}$$

for all x satisfying $(0.0\underline{2})_3 = \tfrac{1}{3} \leq x \leq \tfrac{2}{3} = (0.2\underline{0})_3$. Also,

$$\varphi(x) = \tfrac{1}{4} \quad \text{for} \quad \tfrac{1}{9} \leq x \leq \tfrac{2}{9}; \qquad \varphi(x) = \tfrac{3}{4} \quad \text{for} \quad \tfrac{7}{9} \leq x \leq \tfrac{8}{9}; \quad \ldots.$$

This extended function, which we will continue to denote by the letter φ, is evidently a monotone nondecreasing function mapping I onto I and does not have any jump discontinuities, since every value of I is taken on at least once. Therefore the extended function φ is continuous at every point of I. We also note that the derivative $\varphi'(x) = 0$ for all points $x \in I - F$, since φ is constant on some

neighborhood of such a point. This extended φ is often called the
Lebesgue singular function. Since the Lebesgue measure of the
set F is 0, we see that the derivative of φ exists almost everywhere
and is equal to 0. However, φ is far from being a constant function,
since it maps I onto the set I.

We now define ψ on I to the closed interval $[0, 2]$ by

$$\psi(x) := x + \varphi(x).$$

It is clear that ψ is a monotone nondecreasing function, since it is the
sum of two such functions. In fact, ψ is strictly increasing on I, so
that ψ is a one-one map of I *onto* the closed interval $[0, 2]$. Moreover,
since ψ is the sum of two continuous functions, it is continuous on
I. It follows that the inverse function ψ^{-1}, which maps $[0, 2]$ onto
I, is also continuous. Therefore, ψ is a homeomorphism (that is, a
continuous function whose inverse is also continuous) of I and $[0, 2]$,
and ψ has the property that if B is a Borel set on R, then both $\psi(B)$
and $\psi^{-1}(B)$ are Borel sets.

Since φ is constant on each of the middle third sets in $I - F$, we
see that ψ maps such a middle third set into an interval of the same
length. Consequently,

$$m\big(\psi(I - F)\big) = m(I - F) = 1,$$

and since $m\big([0, 2]\big) = 2$, it follows from the fact that $[0, 2] = \psi(F) \cup$
$\psi(I - F)$ and $\psi(F) \cap \psi(I - F) = \emptyset$, that we have

$$2 = m\big(\psi(F)\big) + m\big(\psi(I - F)\big).$$

Consequently, we must have $m\big(\psi(F)\big) = 1$. We conclude that the
homeomorphism ψ maps the set F, which has Lebesgue measure 0,
to a set with Lebesgue measure equal to 1.

Since $\psi(F)$ has positive measure, we know from Theorem 17.7
that it contains a set W that is not Lebesgue measurable. Then the
set $W_1 := \psi^{-1}(W)$ is a subset of F and hence is a Lebesgue null
set; consequently, W_1 is a Lebesgue measurable set. However, W_1
cannot be a Borel set, since if it were, then $W = \psi(W_1)$ would also

be a Borel set, and hence a Lebesgue measurable set.   But this is contrary to the choice of W as a nonmeasurable set.

We state this result formally.

17.11 THEOREM.   *There exist Lebesgue measurable subsets of* R *that are not Borel sets.*

We observe that we have also given an example of a Lebesgue measurable set W whose image under a strictly monotone homeomorphism (and hence Borel measurable function) ψ is not Lebesgue measurable. *Thus, although a homeomorphism always maps Borel sets to Borel sets, it may map a Lebesgue measurable set to a nonmeasurable set.*

References

1. Bartle, R. G., *The Elements of Real Analysis*, Second edition, John Wiley and Sons, New York, 1976.

2. Bartle, R. G., and D. R. Sherbert, *Introduction to Real Analysis*, Second edition, John Wiley and Sons, New York, 1992.

3. Bourbaki, N., *Éléments de Mathématique, Livre VI, Intégration*, Chapters 1-4 (1952), Second edition (1965); Chapter 5 (1956), Second edition (1965); Chapter 6 (1959); Chapters 7-8 (1963); Chapter 9 (1969), Hermann et Cie, Paris.

4. Dudley, R. M., *Real Analysis and Probability*, Wadsworth and Brooks/Cole, Pacific Grove, 1989.

5. Dunford, N., and J. T. Schwartz, *Linear Operators, Part I*, Interscience Pub. Co., New York, 1958. Wiley Classics Library.

6. Evans, L. C., and R. F. Gariepy, *Measure Theory and Fine Properties of Functions*, CRC Press, Boca Raton, 1992.

7. George, Claude, *Exercises et Probèmes d'Intégration*, Gauthier-Villars, Paris, 1980. English edition, Springer-Verlag, New York, 1984.

8. Halmos, P. R., *Measure Theory*, D. Van Nostrand, New York, 1950. Second edition, Springer-Verlag, New York, 1988.

175

9. Hawkins, Thomas, *Lebesgue's Theory of Integration: Its Origins and Development*, The University of Wisconsin Press, Madison, 1970.

10. Munroe, M. E., *Introduction to Measure and Integration*, Addison-Wesley, Reading, 1953.

11. Oxtoby, J. C., *Measure and Category*, Springer-Verlag, New York, 1971.

12. Shilov, G. E., and B. L. Gurevich, *Integral, Measure and Derivative: A Unified Approach*, English Edition, Dover Pub. Inc., New York, 1977.

13. Solovay, R. M., *A model of set theory in which every set of reals is Lebesgue measurable*, Annals of Math. (2) 92 (1970), 1 – 56.

14. Taylor, A. E., *General Theory of Functions and Integration*, Blaisdell Pub. Co., New York, 1965. Reprinted, Dover Pub. Inc.

Index

Absolute continuity, 35, 84
Absolute integrability, 42
Additivity, countable, 4, 19
Algebra, Borel, 7
  extended Borel, 8
  of sets, 96
Almost everywhere, 22
Almost everywhere convergence, 22, 65 ff.
Almost uniform convergence, 72 ff.
Archimedes, 1

Banach space, 60
Borel algebra, 7
  measure, 20, 104
  measurability, 9
  set, 7, 151

Cantor set, 153, 171
  singular function, 173
Carathéodory, C., 100, 141
  Extension Theorem, 101, 142
Cauchy sequence in L_p, 58
Cauchy–Bunyakovskiĭ–Schwarz
  Inequality, 57
Cell, 127, 129
Characteristic function, 3, 9
Charge, 23
Closure, 128, 129
Combinations of measurable
  functions, 9
Complete measurable
  space, 103, 152
Complete normed space, 58
Completeness Theorem, 59

Completion of a measurable
  space, 25
  of Borel measure, 110
  of a measure, 25
Complex-valued functions, 13, 49
Continuity, absolute, 35, 84
Convergence, almost
  everywhere, 22, 65
  almost uniform, 72 ff.
  dominated, 44, 75
  in L_p, 58, 66 ff.
  in mean, 66 ff.
  in measure, 68 ff.
  pointwise, 65
  uniform, 65
Countably additive, 19, 139
  subadditive, 100, 135
Counting measure, 20, 141
Cube, 129

De Morgan's laws, 7
Decomposition of
  measures, 80-95
Diameter of a set, 134
Difference set, 165
Difference, symmetric, 160
Dominated Convergence
  Theorem, 44, 75

Egoroff's Theorem, 74
Endpoint of an interval, 2, 127
Equivalence of functions, 54
  rational, 166
Essentially bounded function, 60
Eudoxos, 1
Extension of measures, 96-112

F_σ-set, 150
Fatou's Lemma, 33, 39
Field, 96
Fréchet, M., 4
Fubini's Theorem, 119
Function, characteristic, 3, 9
    complex-valued, 13
    continuous, 9
    essentially bounded, 60
    imaginary part of, 14
    integrable, 41
    measurable, 8, 10
    measure, 19
    monotone, 9
    negative part of, 10
    positive part of, 10
    real part of, 14
    simple, 3, 27
    step, 3, 38
    summable, 41
Functional, represen-
    tation of, 90 ff., 106

G_δ-set, 150

Hahn, H., 103
Hahn Decomposition, 81-82
    Extension Theorem, 103
Half-closed, 127
Half-open, 127

Inequality, Cauchy–Bunya-
    kovskiĭ–Schwarz, 57
    Hölder, 56
    Minkowski, 57
Infimum, 5
Inner measure, 164
Integrable function, 41
Integral, indefinite, 42
    of complex-valued functions, 49
    of nonnegative measurable
        functions, 30
    of simple functions, 28
    of real-valued functions, 41
    Riemann, 3, 38
Interior, 128, 129
Interval, 2, 128
Invariance under translation, 130

Joichi, J. T., 78-79
Jordan Decomposition, 83

Lebesgue, H., 1
Lebesgue Decomposition, 88
    Dominated Convergence, 44
    measure, 20, 104, 144
    σ-algebra, 144
    spaces, 52-64
Lebesgue–Stieltjes measure, 105
Leibniz, G. W., 1
Length, of an interval, 3, 96, 128
Levi, B., 31
Limit, of a sequence of extended
    real numbers, 5
    of a sequence of sets, 16
Linear functional, 89 ff., 106

Mean convergence, 66 ff.
Measurable function, 8, 10
    set, 7
    space, 6
Measure, 19, 140
    Borel, 20, 104
    Borel–Stieltjes, 20, 105
    construction of, 96-112
    convergence in, 68 ff.
    counting, 20
    finite, 20
    generated by a monotone
        function, 20, 105
    inner, 164
    Lebesgue, 20, 105, 144
    Lebesgue–Stieltjes, 105
    outer, 99, 132
    σ-finite, 20
    singular, 88
    space, 22
Minkowski Inequality, 57
Monotone class, 17, 18, 116
    Class Lemma, 116
    Convergence Theorem, 31
    function, 9

Negative part, 10
    set, 80
    variation, 82
Newton, I., 1
Non-Borel, 154, 167 ff.
Nonmeasurable, 104, 171 ff.
Norm, 52
    of a linear functional, 89
Nowhere convergence, 68
Null set, 80, 151

Outer measure, 99, 132

Pointwise convergence, 65
Positive part, 10
    set, 80
    variation, 82
Product of two measures, 115
Pseudo-norm, 52

Radon–Nikodým derivative, 87, 94
    Theorem of, 85, 94, 95
Rational equivalence, 166
Rectangle, 113
Representation theorems for
    functionals, 90, 106
Riemann, B., 1
Riemann integral, 3, 38
Riesz, F., 58, 69, 90, 106
Riesz Representation Theorems,
    on C, 106 ff.
    on L_1, 90 ff.
    on L_p, 91 ff.

Schwarz Inequality, 57
Section, 115
Semi-norm, 52
Set, Borel, 7, 151
    Cantor, 7, 151
    closed, 150
    difference, 165
    Lebesgue, 144
    measurable, 100, 144
    negative, 80
    nonmeasurable, 104, 167 ff.
    null, 80, 151
    open, 149
    positive, 80
    ubiquitous, 169
σ-algebra, 6, 140
    generated by a family, 7
σ-field, 6
σ-finite measure, 21
Simple function, 3, 27
Singular function, 173
    measure, 88

Space, measurable, 6
    measure, 22
    product, 113
Standard representation, 27
Step function, 3, 38
Stieltjes–Lebesgue measure, 105
Subadditivity, 100, 135, 139
Supremum, 5
Symmetric difference, 160

Theorems, Approximation of
    measurable sets, 155 ff.
    Carathéodory
    Extension, 101, 142
    Completeness, 59
    Dominated Convergence, 44
    Egoroff, 74
    Fatou, 33
    Fubini, 119
    Hahn Decomposition, 81
    Hahn Extension, 103
    Jordan Decomposition, 83
    Lebesgue Decomposition, 88
    Lebesgue Dominated, 44
    Monotone Class, 116
    Monotone Convergence, 31
    Product Measure, 114
    Radon–Nikodým, 85
    Riesz–Fischer, 59
    Riesz Representation, 90, 91, 106
    Riesz Subsequence, 69
    Tonelli, 118
    Vitali Convergence, 76
Total variation, 83
Translation invariance, 130, 137, 153
Truncation, 16

Ubiquitous set, 169
Uniform convergence, 65 ff.
Unit measure, 20

Variation, 25, 82-83
Vitali Convergence Theorem, 76
Vitali's set, 167
Volume, 129